MANAGING RISK IN CONSTRUCTION PROJECTS

Second Edition

Nigel J. Smith
Professor of Project & Transport Infrastructure Management
School of Civil Engineering
University of Leeds

Tony Merna
Civil & Construction Engineering
School of Mechanical, Aerospace and Civil Engineering
University of Manchester

Paul Jobling
Project Director Risk Management
Senior Professional Associate
Parsons Brinckerhoff

Blackwell
Publishing

© 2006 N. J. Smith
© 1999 By Blackwell Science Ltd
Editorial offices:
Blackwell Publishing Ltd, 9600 Garsington Road, Oxford OX4 2DQ, UK
 Tel: +44 (0) 1865 776868
Blackwell Publishing Inc., 350 Main Street, Malden, MA 02148-5020, USA
 Tel: +1 781 388 8250
Blackwell Science Asia Pty Ltd, 550 Swanston Street, Carlton, Victoria 3053, Australia
 Tel: +61 (0)3 8359 1011

First published 1999
Transferred to digital print 2003
Second edition published 2006

ISBN-13: 978-1-4051-3012-7
ISBN-10: 1-4051-3012-1

Library of Congress Cataloging-in-Publication Data
Smith, Nigel J.
   Managing risk: in construction projects/Nigel J. Smith, Tony Merna, Paul
Jobling.–2nd ed.
    p. cm.
 Includes bibliographical references and index.
 ISBN-13: 978-1-4051-3012-7 (alk. paper)
 ISBN-10: 1-4051-3012-1 (alk. paper)
1. Building–Superintendence. 2. Building–Safety measures. 3. Construction
industry–Management. 4. Risk assessment. I. Merna, Tony. II. Jobling, Paul 1955- III. Title.

TH438.S54 2006
690′.068–dc22   2005010452

A catalogue record for this title is available from the British Library

Set in 10/13 pt Times NR
by Newgen Imaging Systems (P) Ltd, Chennai, India
Printed and bound in India
by Replika Press, Pvt Ltd., Kundli

The publisher's policy is to use permanent paper from mills that operate a sustainable forestry
policy, and which has been manufactured from pulp processed using acid-free and elementary
chlorine-free practices. Furthermore, the publisher ensures that the text paper and cover board
used have met acceptable environmental accreditation standards.

For further information on Blackwell Publishing, visit our website:
www.thatconstructionsite.com

Contents

Preface

Those of you wanting the answer to the problems of risk management might think of turning straight to the final chapter. Indeed, there you will find a summation of how risk management methods can empower the decision-making of the project manager. However, it is only a thorough understanding of the various concepts involved that can provide the real basis on which to make effective decisions.

The essence of the guidance is based on the interaction of concepts, user requirements and specific projects, and it is by obtaining a greater knowledge of the inherent nature of the project that improvements in performance can be found. Hence by examining the guidance in this context, the reader will be able to gain the maximum benefit from this book. The authors doubt many people will read this book from cover to cover but if key sections of the text serve to enhance understanding and to facilitate more effective project management then it will have achieved its purpose.

The second edition of this book has been extended to include the input of the Turnbull Report and to introduce the concept of corporate, strategic business project level risk. Nevertheless, the basic concept of risk management as a process for making better decisions under conditions of uncertainty remains.

This book is not intended as a definitive monograph on risk but as a guide for practitioners having to manage real projects. The authors have assembled a strong team of practitioners and leading academics and it is the blend of theory and practice which is the real message of this work.

Authors Biographies

Paul Jobling BSc, MSc, CEng, MICE, MAPM is a Senior Professional Associate of Parsons Brinckerhoff and Project Director for Project Risk Management. He has worked in the field of project and programme management since 1976. He was a member of the research team that produced the *Guide to Risk Management in Construction* published in 1986. At Eurotunnel he worked in the project control team developing procedures for risk analysis and contingency fund management. Further risk management and analysis work has included the Channel Tunnel Rail Link, major nuclear decommissioning programmes and several major rail programmes including the West Coast Route Modernisation, Train Protection and Warning System, Southern Region New Trains Programme and the European Rail Traffic Management System. Paul was a member of the working party responsible for the production of the *Project Risk Analysis and Management Guide* published in 1997 by the Association for Project Management, and a member of the review team for the revised edition published in 2004.

Anthony Merna BA, MPhil, PhD, CEng, MICE, MAPM, MIQA is senior partner of Oriel Group Practice, a multi-disciplinary research consultancy based in Manchester and a lecturer in the School of Mechanical, Aerospace and Civil Engineering (MACE), at the University of Manchester. He currently teaches risk management to MSc and MBA students at a number of UK and overseas institutions, and supervises MPhil and PhD students researching in risk management. He advises numerous organisations on the application of risk management at corporate, strategic business and projects levels.

Nigel J. Smith BSc, MSc, PhD, CEng, FICE, MAPM is Professor of Project & Transport Infrastructure Management and Head of School, at the School of Civil Engineering, University of Leeds. After graduating from the University of Birmingham, he gained practical experience with Wimpey, North East Road Construction Unit and the Department of

Transport. Since returning to academia he has researched and published widely in the field of risk management and regularly teaches on MSc and CPD Risk Management Courses. He has presented papers on risk at many international conferences including Trondheim, Budapest, Florence, Moscow, Bonn and Brisbane.

Acknowledgements

I am particularly grateful to my co-authors in this second edition, Tony Merna and Paul Jobling, for helping to update, modify and improve the existing text blending theory and practice. I would also like to acknowledge the assistance of Tony Merna Jr and Douglas Lamb for their expertise in drafting new sections of the text. In addition I would like to recognise the work of all the original authors of the first edition, namely Dr Chris Adams, Dr Denise Bower, Mr Otto Husby and Ms Trina Norris.

I would like to express my thanks once again to Ms Sally Mortimer of the School of Civil Engineering for processing, checking and questioning the book text and for her help with all aspects of the administration of the writing and editing process.

Nevertheless, as was the case with the first edition, I take the responsibility for any residual risks associated with any errors in the book.

Professor Nigel J Smith

Chapter 1

Projects and Risk

This book concentrates on aspects of risk management and also clarifies the practical procedures for undertaking and utilising decisions. Risk management is beset by a dark cloak of technology, definitions and methodologies, often maintained by analysts and specialist consultants, which contributes to the unnecessary mystique and lack of understanding of the approach. It discusses a number of general concepts including projects, project phases and risk attitude before introducing a number of risk management techniques. The book concludes with some brief case studies and guidance on good practice.

This book offers for the first time – in the opinion of the authors – the distilled knowledge of over a hundred man-years of project experience in working on aspects of project risk management and contains information which most of us would have liked to have had – had it been available and collated. To all students and practitioners using this book, follow known procedures as outlined in the book, avoid *short-cuts* and remember to keep records of everything you model, simulate or assume.

1.1 Construction projects

Change is inherent in construction work. For years, industry has had a very poor reputation for coping with the adverse effects of change, with many projects failing to meet deadlines and cost and quality targets. This is not too surprising considering that there are no known *perfect* engineers, anymore than there are *perfect designs* or that the forces of nature behave in a *perfectly* predictable way. Change cannot be eliminated, but by applying the principles of risk management, engineers are able to improve the effective management of this change.

Change is normally regarded in terms of its adverse effects on project cost estimates and programmes. In extreme cases, the risk of these time and cost overruns can invalidate the economic case for a project, turning

1

a potentially profitable investment into a loss-making venture. A risk event implies that there is a range of outcomes for that event which could be both more and less favourable than the most likely outcome, and that each outcome within the range has a probability of occurrence. The accumulation, or combinations of risks can be termed *project risk*. This will usually be calculated using a simulation model (see Chapter 7). It is important to try to capture all the potential risks to the project even if they are not strictly events or a calculation of project risk.

In construction projects each of the three primary targets of cost, time and quality will be likely to be subject to risk and uncertainty. It follows that a realistic estimate is one which makes appropriate allowances for all those risks and uncertainties which can be anticipated from experience and foresight. Project managers should undertake or propose actions which eliminate the risks before they occur, or reduce the effects of risk or uncertainty and make provision for them if they occur when this is possible and cost effective. It is vital to recognise the root causes of risks, and not to consider risks as events that occur almost at random. Risks can frequently be avoided if their root causes are identified and managed before the adverse consequence – the risk event – occurs. They should also ensure that the remaining risks are allocated to the parties in a manner which is likely to optimise project performance.

To achieve these aims it is suggested that a systematic approach is followed: to identify the risk sources, to quantify their effects (risk assessment and analysis), to develop management responses to risk and finally to provide for residual risk in the project estimates. These four stages comprise the core of the process of risk management. Risk management can be one of the most creative tasks of project management.

The benefits of risk management can be summarised as follows:

- ❏ project issues are clarified, understood and considered from the start;
- ❏ decisions are supported by thorough analysis;
- ❏ the definition and structure of the project are continually monitored;
- ❏ clearer understanding of specific risks associated with a project;
- ❏ build-up of historical data to assist future risk management procedures.

1.2 Decision making

Risk management is a particular form of decision making within project management, which is itself the topic of many textbooks and papers. Risk management is not about predicting the future. It is about understanding your project and making a better decision with regard to the management of your project, tomorrow. Sometimes that decision may be to abandon the project. If that is the correct outcome which saves various parties from wasting time, money and skilled human resources, then the need for a rational, repeatable, justifiable risk methodology and risk interpretation is paramount. Nevertheless, the precise boundaries between decision making and the aspects of other problem-solving methodologies have always been difficult to establish.

In essence, decisions are made against a predetermined set of objectives, rules and/or priorities based upon knowledge, data and information relevant to the issue although too often this is not the case. Frequently decisions are ill-founded, not based on a logical assessment of project-specific criteria and lead to difficulties later. It is not always possible to have conditions of total certainty; indeed in risk management it is most likely that a considerable amount of uncertainty about the construction project exists at this stage.

The terms risk and uncertainty can be used in different ways. The word risk originated from the French word *risqué*, and began to appear in England, in its anglicised form, around 1830, when it was used in insurance transactions. Risk can be, and has been, defined in many ways and assessed in terms of fatalities and injuries, in terms of probability of reliability, in terms of a sample of a population or in terms of the likely effects on a project. All these methodologies are valid and particular industries or sectors have chosen to adopt particular measures as their standard approach. As this book concentrates on engineering projects, risk is defined in the project context, and broadly follows the guidelines and terminology adopted by the British Standard on Project Management BS 6079, The Association for Project Management Body of Knowledge, The Association for Project Management Project Risk Analysis and Management Guide, the Institution of Civil Engineers and the Faculty of Actuaries Risk Analysis and Management for Projects Guide and the HM Treasury, Central Unit on Procurement Guide on Risk Assessment.

A number of authors state that uncertainty should be considered as separate from risk because the two terms are distinctly different. Uncertainty can be regarded as the chance occurrence of some event where the probability distribution is genuinely not known. This means that uncertainty relates to the occurrence of an event about which little is known, except the fact that it may occur. Those who distinguish uncertainty from risk define risk as being where the outcome of a event, or each set of possible outcomes, can be predicted on the basis of statistical probability. This understanding of risk implies that there is some knowledge about a risk as a discrete event or a combination of circumstances, as opposed to an uncertainty about which there is no knowledge. In most cases, project risks can be identified from experience gained by working on similar projects.

Risks fall into three categories; namely known risks, known unknowns and unknown unknowns. Known risks include minor variations in productivity and swings in material costs. These occur frequently and are an inevitable feature of all construction projects. Known unknowns are the risk events whose occurrence is predictable or foreseeable. Either their probability of occurrence or their likely effect is known. Unknown unknowns are those events whose probabilities of occurrence and effect are not foreseeable by even the most experienced staff. These are usually considered as force-majeure.

In some situations the term risk does not necessarily refer to the chance of bad consequences, it can also refer to the possibility of good consequences, therefore, it is important that a definition of risk must include some reference to this point. Risk and uncertainty have been defined as:

❏ *risk* exists when a decision is expressed in terms of a range of possible outcomes and when known probabilities can be attached to the outcomes;
❏ *uncertainty* exists when there is more than one possible outcome of a course of action but the probability of each outcome is not known (frequently termed estimating uncertainty).

A particular type of decision making is needed in risk management. Consider Figure 1.1 which compares the probability of occurrence of an event compared with its impact on the construction project. Events with a low impact are not serious and can be divided into the elements of trivial and expected. For the high impact and low probability, these

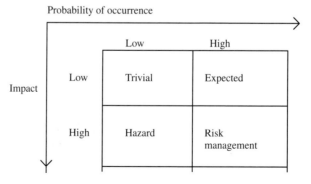

Figure 1.1 Classification of risk sources.

events are a hazard which could arise but are too remote to be considered. For example, there is a finite probability that parts from an old satellite might re-enter the atmosphere and crash on any building project in the United Kingdom, but very few buildings need to be designed to withstand that event. In project management however, high impact risks should not be ignored even if their probability is low. Fallback and response plans should be put in place even if the financial impact is too large to be covered by contingencies. The use of risk management is to identify, assess and manage those events with both a high input and a high probability of occurrence.

1.3 Risk management strategy

Most commonly, the client, the project owner (e.g. companies, organisations, etc.) has an overall risk management strategy and policy included in the strategic documents and quality management system. Main issues concerning project owner risk strategy are risk ownership (which party owns the risk; risk exposure and transfer) and risk financing (how to include and use budget risk allowance or contingency). The client's risk management policy includes the risk management procedures or guidelines, responsibilities and reporting.

Both client (employer, promoter) and contractor are concerned with the magnitude and pattern of their investment and the associated risk. They desire to exert control over the activities which contribute to their investment. This type of risk is now covered by the term corporate and project governance (see Chapter 10).

There are two significant axioms of control: (1) control can be exercised only over future events and (2) effective control necessitates prediction of

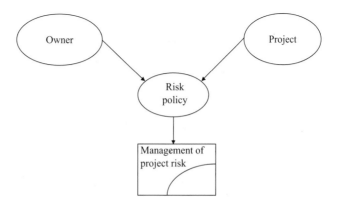

Figure 1.2 Risk management strategy.

the effects of change. The past is relevant only so far as past performance or events can influence our predictions of the future. The scope for control diminishes as the project proceeds. There are two key events at which control can be exercised; (1) sanction commitment to a project of particular characteristics and (2) contract award commitment to contractors and major cost expenditure. It should be noted that there will also be opportunity to influence even if direct control cannot be exercised.

It follows that prior to these two commitments clients have great opportunity for control. They make decisions to define the organisation and procedures required for the execution of a project. These decisions affect the responsibilities of the parties; they influence the control of design, construction, commissioning, change and risk; hence they affect cost, time and quality.

1.4 Project planning

The control of time cannot be effected in isolation from resources and costs. Project planning methods should be utilised to communicate to all parties in a project, to identify sequences of activities and to draw attention to potential problem areas. The successful realisation of a project will depend greatly on careful planning and continuous monitoring and updating. The activities of designers, manufacturers, suppliers, contractors and all their resources must be organised and integrated to meet the objectives set by the client and/or the contractor. In most cases the programme will form the basis of the plan.

Sequences of activities will be defined and linked on a timescale to ensure that priorities are identified and that efficient use is made of expensive and/or scarce resources. Remember, however, that because of the uncertainty it should be expected that the plan will change. It must therefore be updated quickly and regularly if it is to remain as a guide to the most efficient way of completing the project. The programme should therefore be simple, so that updating is straightforward and does not demand the feedback of large amounts of data, and flexible, so that all alternative courses of action are obvious.

The purposes of planning are therefore to persuade people to perform tasks before they delay the operations of other groups of people, and in such a sequence that the best use is made of available resources and to provide a framework for decision making in the event of change. It is difficult to enforce a plan which is conceived in isolation, and it is, therefore, essential to involve the individuals and organisations responsible for the activities or operations as the plan is developed.

In developing a plan which is to be used for purposes of control, it is vital to distinguish between different categories of change and to fully instigate the monitoring and formal aspect of the project. Typically, the main categories are: adapted, fixed (e.g. for mobilisation); time related (e.g. for resources and overheads) and quantity-proportional (e.g. for materials). Their relative importance will differ with the project and it is interesting to note the importance of time-related costs and the implications of delay in plant-intensive construction projects.

Project management information systems (PMISs) should forecast the outcome of a project in terms related to achievement of its objectives. Integrated cost models link time with money. They provide project managers with forecasts to completion in terms of cost, time resource usage and cash flow. Decisions about future actions can be made with the best available forecasts in these terms. Cost models also help to overcome an implementation gap between monitoring systems and the manager's action. Risk management software (RMS) is the term used to denote a specialist software, which can be used to apply one of the many risk assessment methodologies.

Project control and information systems should be conceived and adopted to suit the needs of a particular project. The project should not be forced to fit the control system; rather the control system should fit the project. Software needs to be selected with due regard to the resources that will be required to operate it and its data requirements.

1.5 Summary

All projects are subject to risk. The world is in a state of constant change and survival relies on the ability to adapt to changes. Unfortunately, many project managers have not yet realised that there is a need to include project risk as a key process.

It is a well known fact that managing risk has two major objectives: to avoid the downside risks and to exploit opportunities. Experiences so far show that the risk avoidance part of the risk management philosophy has attracted too much management attention, while the potential opportunities have been neglected.

The risk avoidance strategy helps you to secure your project objectives, which for many organisations is a giant step ahead and may be the single biggest opportunity. However, the major leaps in project cost and time reduction are results of innovative thinking with focus on exploring opportunities by challenging the risks. The trend today is to establish ambitious goals, to seek for new technological solutions and concepts and to look for effective ways of organising and managing projects.

The difference between project success and disaster is of course more complex than managing or not managing the risk, but it appears that the number of successful projects would have been far higher if more companies had included risk as an integral part of their project management.

The following chapters present a framework against which a practicable and rational approach to the process of managing risk in construction projects can be developed.

Chapter 2
The Project Environment

2.1 Projects

Projects do not exist in isolation. They are initiated to fulfil a need or exploit an opportunity. The needs and opportunities exist before the project. They are products of the world at large. Projects are therefore heavily influenced by external factors and they also influence the world outside them to an extent that is largely, but not entirely, dependent upon the size of the project.

These external factors can be termed the project environment. Other names are also given to it such as the project world. Perhaps the single most important influence on any project is whether or not it is carried out by the public or private sectors. Public sector projects are those undertaken by central and local government whereas private sector projects are those undertaken by individual companies or consortia which are usually entirely privately owned. The aims and objectives of these two sectors are different and projects are undertaken by them for different reasons. The main aim, if not the sole aim, of projects undertaken in the private sector is to make a profit, whereas for projects undertaken by the public sector it is whether the project provides a public service and is also of benefit to the community.

In the United Kingdom in recent years, however, this distinction has become blurred. The increasing burden upon the state of large commitments, including publicly-owned enterprises, coupled with significant increases in funding costs, has meant that increasingly the public sector is looking to the private sector to finance projects. This has led to the trend in recent years for projects to be procured under the Design–Build–Finance–Operate (DBFO) alternatively known as Build–Own–Operate–Transfer (BOOT) or Build–Operate–Transfer (BOT). In the United Kingdom, all projects of these types are now known as public-private partnerships (PPP) or private finance initiative (PFI) projects. This is intended to reduce public expenditure on both the capital and running costs by transferring them to the private sector. This is perceived to have the dual advantage of: first, reducing the requirements for public expenditure on capital projects; and second, producing projects which can be operated more efficiently thereby reducing the requirement for public funding of the operating costs. It is also perceived that such projects have the further advantage of reducing capital costs by reducing the incidence of overspecification and overdesign, and by reducing conflict between the various parties to the project by creating a single entity, which combines the consultant, contractors and operators. In such projects, public need is serviced only where it can generate a profit during the operation phase. In fact, audits have begun to question the value for money and long-term benefits of this approach to procuring projects and concerns have been raised about the long-term commitment the public sector now has to continue payment for the services provided.

Generally speaking, in publicly funded projects, the government or local authorities have taken many of the risks. This has been true in the past of private sector projects too. Recently however, private companies and consortia have sought to transfer more of the risk for the design and construction of their project to the consultants and contractors who design and construct them. This has come about because the private sector is increasingly concerned at the incidence of delayed completion and increased costs brought about by the more traditional ways of procuring and implementing capital projects. The consequence has been the development of a number of alternative types of procurement strategy, the most common of which is some form of turnkey contract where one entity is responsible for both the design and the construction. This is believed to put greater responsibility upon that party and remove some of the potential conflict, thereby reducing the incidence of cost overspend and programme delay.

This is an example of the transfer of risk from the owner to a contractor. Nevertheless, in such arrangements, the owner would retain the risk of the viability of the project and that of the operating and maintenance costs. Those projects that are being carried out under PFI, seek to transfer these risks to the private sector by combining the designer, contractor and operator into a concessionaire organisation. The latter is responsible for raising the capital, managing the project and is then responsible for the operation and maintenance of the asset to a predetermined specification for which it receives some form of fee as income from the public sector or from users, for providing the service.

The importance of distinguishing between these types of project is that they fundamentally determine the attitudes towards risk assessment, risk transfer and risk management which must be adopted in the

initiation and implementation of such projects. It should be noted however that the National Audit Office (NAO) now questions whether or not
risk transfer has really taken place from the public sector to the private
consortia.

2.2 The project constitution

The next most important influence on a project is its governance
arrangements or constitution, namely who the members of the owners (client/promoter/concessionaire) organisation are, other stakeholders
and third parties such as government or statutory authorities; what their
relationships are, how the relationships are structured and where the
authority lies. The owner may be a single entity, a private company or a
government department, or it may be a group of private or public organisations, which combine in some form of partnership or consortium to
promote the project. Clearly, a single entity provides a simpler constitution than a multientity owner. Indeed, for a single entity working in
a clearly defined business, such as, a supermarket chain building a new
store, the term constitution is probably unnecessary and the term organisation is adequate. However, for the multiparty owner it is essential
that it be recognised that the term organisation is inadequate to define
the context within which the project will be executed. All projects will
have some form of organisation – which may be quite simple – but it
is the way in which the owner/promoter organisation is put together or
constituted, which is important. For example, oil production facilities in
the North Sea require the combined resources of several oil companies
each of whom then takes a stake in the revenues. One of the companies
is given the responsibility for managing the project on behalf of the other
stakeholders. Contrast this with the way in which the constitution of
the owner/promoter/concessionaire for the Channel Tunnel evolved. The
original intention was that a concessionaire consisting of construction
contractors, designers and bankers would create an operating company
to operate and maintain the project, constructed, designed and funded
by the consortium. However, shortly after being awarded the concession,
the consortium split into its constituent parts creating in the process a new
entity that came to be known as Eurotunnel the concessionaire. The banks
became purely funding institutions and the contractors formed a consortium to design and construct the project. The governments also continued
to influence the project by way of the inter-governmental commission who
had overriding responsibility for ensuring that the project met the concession specification. Lessons learned from the Channel Tunnel led to

a different approach by the successful bidder for the Channel Tunnel rail link concession, which is described in Chapter 11.

These examples illustrate the different types of constitution, which projects can have: a simple constitution, a more complex but nevertheless clearly defined constitution or a complex constitution with split responsibilities and ill-defined authority.

The constitution is important because the owner is responsible for making the key decisions, and any constraints on his ability to do so must be clearly identified and understood. This is essential because the speed and decisiveness, which the owner brings to decision making, is crucial to the success of projects. The more complex the constitution – and the less clearly defined the hierarchy – the slower will be the speed with which decisions are made which could result in delays to the project. If decisions lack certainty, confusion will result, and there will be a need to make further decisions to clarify earlier statements. This will result in changes to the project that will usually have adverse impacts. The later the change the greater the impact. Delays to the programme are the most obvious consequence but inevitably these also lead to increased costs and possibly to changes to the functionality and quality.

It is probably true to say that public projects usually have the most complex constitutions when the Treasury, at least one government department and probably more, have interests in a project and influence over its conception, design and execution. The British Library was an example and became notorious for huge delays and cost overruns. In the case of the private sector, complex constitutions may also be common, but the importance of achieving agreement of the project's objectives, the need for a clear hierarchy and single point responsibility and certainty are better understood as essential to the success of projects; hence the adoption of the constitution for the North Sea projects as described above.

Of course, a simple constitution on its own does not guarantee success. There is a multitude of other factors to consider, many of which will be discussed in this book, but without a constitution created with the express intent of delivering a successful project, the chances of success are greatly reduced.

Another facet to the successful management of projects, that was often ignored in the past but is now being increasingly recognised, is the influence of third parties such as regulatory agencies and single issue pressure groups, most notably the environmental lobby. These groups can wield significant influence and exert significant pressure on the project to the point of forcing major changes, such as the re-routing of highways schemes, or cancellation of waste disposal projects. It is essential that the views of these groups are canvassed, understood and wherever possible accommodated. Management time and effort must be directed towards these organisations otherwise there is the risk of their intervention at a time, which is disadvantageous to the project. The establishment of clear lines of communication and good working relationships are a prerequisite of managing these groups, the ultimate objective of which is to establish a situation of mutual trust and understanding.

These third party influences are part of the project's environment and the project world. If their influence cannot be accommodated by the project in its concept or design, provision must be made in other ways, such as allowing time for public enquiries and contingency budgets for any modifications, which are required as a result.

Decisions concerning the way in which the project is constituted, the roles of the stakeholders; roles and influence of the third parties; the way in which these relationships are structured, by written or by other means; and the channels and frequency of communication, must be considered extremely carefully. The objective must be to arrive at a constitution, which is geared up to the delivery of a successful project, not a constitution that suits the preferred *modus operandi* of the parties, but fails to address the needs of the project.

2.3 Project organisation

Organisation means the way in which the project's implementation team is organised and who the participants are.

Projects can be split in single discipline and multidiscipline. The traditional civil engineering sector has been single discipline, while the building and process engineering sectors have been multidiscipline but as projects have become larger and more complex it is becoming more likely that projects are multidiscipline. For example, many highway projects contain sophisticated traffic signing, information systems and speed cameras. Similarly, the signalling and control systems for rail projects are becoming more sophisticated and expensive. Hence, they are now a larger proportion of the project than they used to be, such that these projects are now clearly multidiscipline.

Single discipline projects

By their nature, these projects are annually undertaken by one project team, frequently staffed by one consultant with a single client and executed by a single contractor. The number of interfaces between individuals and organisations are relatively few and easily managed.

The most complicated relationships exist between the contractor, his sub-contractors and suppliers (i.e. the contractor's supply chain). However, these relationships and the structure required to manage them is relatively simple, although in recent years it has been recognised that significant savings can be made by dedicating effort to managing the supply chain. Hence, these projects usually represent a lower risk than multidiscipline projects, even though they can be large and have high values.

Multidiscipline projects

Despite the greater organisational complexity, projects can be quite small. For example, even quite small buildings may require:

- civil engineering input if the foundations are complex;
- a structural engineer for the building superstructure;
- a building services engineer;
- an architect to lead the team and prepare the overall design;
- a mechanical and engineering (M&E) or a process contractor;
- a fitting out contractor.

Other specialists involved may include telecommunications engineers, lift specialists and cladding specialists. The contractor's organisation may be equally complex with specialist trade sub-contractors for civil works, structural works, brickwork, carpentry, plumbing, installation of services, telecommunications and so on. To complicate matters further, the specialist sub-contractors may also have impact into the design process, for example the sizing of lift shafts and machinery rooms.

Traditional procurement methods often split responsibility in an unrealistic and arbitrary way that cuts across work packages. For example, the overall design of electrical systems and HVAC (heating, ventilation and air conditioning) may be the responsibility of a design consultant; the co-ordination of M&E services may be the responsibility of the main contractor; while the detailed design of the HVAC installation may be the responsibility of a specialist supplier. The structural engineer meanwhile is responsible for the design of the building frame, although detailing may be the responsibility of the fabricator.

Clearly, this type of organisational structure increases the risks and likely results in poor communications, delays and incorrect information leading to claims and disputes. It is for this reason that clients in the building sector – especially developers – have moved to other forms of procurement including design and build, because, though different disciplines are still present, they are all part of a single organisation.

Increasingly large clients are moving towards partnering arrangements with selected suppliers so that the various disciplines are effectively part of one organisation. This is aimed at minimising the risks described above with the objective of improving the chances of successful delivery of projects, on time, to budget and to specification. Partnering has been successful in manufacturing industry and on projects where the extent of risk is similar to those encountered in construction projects.

One final point that should be made is that organisations are collections of people. And research has shown that groups of people are less risk aware than individuals. It is possible therefore that complex organisations are more likely to take risks than smaller less complex organisations.

2.4 Project phases

It has been recognised for some time that projects exhibit a life cycle comprising of a number of discreet stages, which as identified by various authors can range from 2 to 12. The former was related to the development of a product and was divided into two phases – product development and implementation, whereas the latter has been developed by the Royal Institute of British Architects (RIBA). It comprises inception, feasibility, outline proposals, scheme design, detailed design, production information, bills of quantity, tender action, project planning, operations on site, completion and feedback.

In other branches of the construction industry the phases are identified as follows: pre-feasibility, feasibility, design, contract/procurement, implementation, commissioning, handover and operation. Different authors give these phases different names, for example, the pre-feasibility stage can be called the inception stage and the initial feasibility stage, the conception stage or the identification stage. However, the precise terminology used is unimportant. Generically, these life cycles and the phases identified are broadly similar and are identified in Figure 2.1.

Other industries have defined other life cycles with different phases and terminology. Despite the differences in terminology and in the number of phases identified, the essence in all cases is the same. The project is divided into a number of discreet phases each of which has a predetermined purpose and therefore an identifiable scope of work. At the completion of each phase, there is a decision point at which progress to date can be reviewed and forthcoming actions identified. These are frequently termed *gateways*. It is now recognised that foremost amongst the information generated during each phase is an assessment of the project's risks. At each decision point therefore, risk assessment is

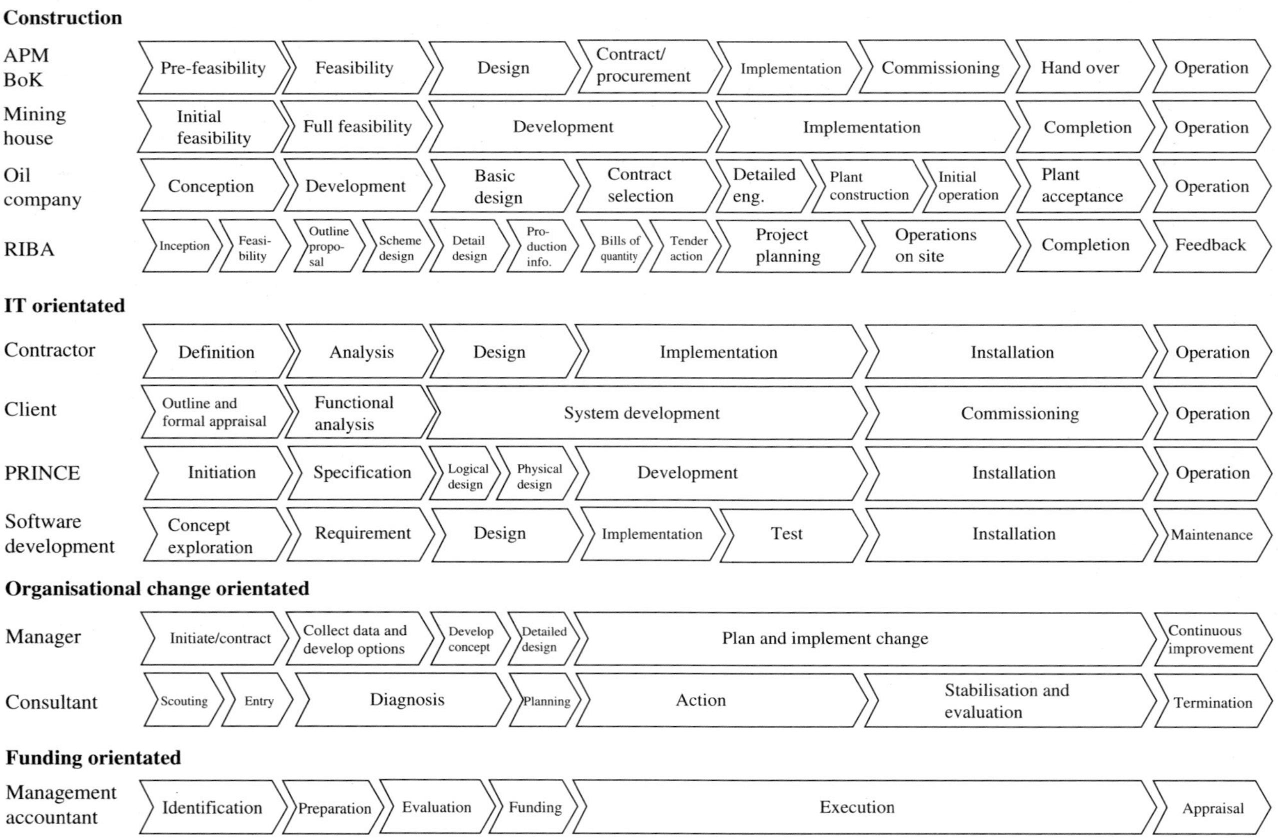

Construction

APM BoK: Pre-feasibility ▸ Feasibility ▸ Design ▸ Contract/procurement ▸ Implementation ▸ Commissioning ▸ Hand over ▸ Operation

Mining house: Initial feasibility ▸ Full feasibility ▸ Development ▸ Implementation ▸ Completion ▸ Operation

Oil company: Conception ▸ Development ▸ Basic design ▸ Contract selection ▸ Detailed eng. ▸ Plant construction ▸ Initial operation ▸ Plant acceptance ▸ Operation

RIBA: Inception ▸ Feasibility ▸ Outline proposal ▸ Scheme design ▸ Detail design ▸ Production info. ▸ Bills of quantity ▸ Tender action ▸ Project planning ▸ Operations on site ▸ Completion ▸ Feedback

IT orientated

Contractor: Definition ▸ Analysis ▸ Design ▸ Implementation ▸ Installation ▸ Operation

Client: Outline and formal appraisal ▸ Functional analysis ▸ System development ▸ Commissioning ▸ Operation

PRINCE: Initiation ▸ Specification ▸ Logical design ▸ Physical design ▸ Development ▸ Installation ▸ Operation

Software development: Concept exploration ▸ Requirement ▸ Design ▸ Implementation ▸ Test ▸ Installation ▸ Maintenance

Organisational change orientated

Manager: Initiate/contract ▸ Collect data and develop options ▸ Develop concept ▸ Detailed design ▸ Plan and implement change ▸ Continuous improvement

Consultant: Scouting ▸ Entry ▸ Diagnosis ▸ Planning ▸ Action ▸ Stabilisation and evaluation ▸ Termination

Funding orientated

Management accountant: Identification ▸ Preparation ▸ Evaluation ▸ Funding ▸ Execution ▸ Appraisal

Figure 2.1 Project phases. From S. McGetrick, The Project Life Cycle, *PROJECT*, June 1996, pp. 13–15, reprinted with kind permission of the author.

a key feature of the decision to proceed to the next phase of the project.

It has been traditional for different parties to be responsible for different phases in the life cycle of the project. Using the APM BoK classification in Figure 2.1, the owner would largely be responsible for the pre-feasibility stages. If a decision to proceed is made then a consultant, usually an architect or engineer, would be appointed to conduct a feasibility study, the objective of which is to compare alternative ways of implementing the project. If at the conclusion of this feasibility stage the project is considered to be viable, taking into the account the costs, benefits and risks, then the next phase of design would also be undertaken by the consultant. After the completion of the design phase, progress can be reviewed and assuming that the project is to proceed, then the next stage would be that of contract/procurement. The next step would be to award the contract for the implementation of the project to the successful tenderer. This implementation stage can be subdivided into several subphases, for example project planning and operations on site. In reality the contractor would be involved in several other subphases such as identifying, preparing documents for and receiving tenders for subcontracted elements of the work. However these are different in character from the main phases identified for the whole life cycle of the project. Following completion of the work there is usually a commissioning procedure in the operation phase.

It is worth noting that from the point of view of the owner and also growing in importance because of the environmental considerations, the decommissioning and disposal of the asset may need to be considered. This is because of the difficulty in dealing with toxic substances which may have been used in the original construction of the asset, such as asbestos or due to the generation of toxic products during the operation.

The main benefit of this approach to the structuring of projects is that a number of key decision points are identified. For example, at the completion of the pre-feasibility stage itself the owner may decide that the project is not worth pursuing. It is also particularly important that exposure to low probability and high impact risks are included in the assessment and not ignored because of the assumption 'they won't happen to us'. The same may apply at the completion of the feasibility stage. Generally if the project is deemed to be feasible then the owner makes a commitment to a detailed design phase, leading up to the procurement stage. Each decision point should be viewed as a gateway with clear criteria for passing through to the next stage or not. The important point here is that the rate of spend will then increase and therefore the decision to move on to the detailed design phase is a major commitment from the owner. The largest commitments are made during the contract procurement stage. First, by the tendering contractors who by submitting a bid undertake to carry out the work if the bid is accepted. The second commitment at this stage is by the owner when a bid is accepted and a contract is entered into, which obliges him to proceed with the implementation. This allocation of risk is defined in many of the standard forms of contract and the decision regarding the form of contract is a crucial decision in the management of risk.

2.5 Effect of project phase on risk

A project is divided into a number of separate phases. At the end of each phase an appraisal can be made and assessment of the risk involved in proceeding with the project. The management of risk is therefore a continuous process and should span all the phases of the project. Since project risks are dynamic, that is to say they can change continuously, a risk assessment must be carried out at the end of each phase prior to proceeding to the next phase. In fact, active management of risk must continue between the review points until the project is complete.

Risks may also change during a phase. Should this be significant then a complete re-appraisal may need to be performed. On long duration projects where the phases themselves may span several months or even years, regular risk assessments and updates must be carried out. This is an essential prerequisite of efficient management and effective decision making.

In addition, to the parties involved changing as the phases change, the nature of the risk itself changes. At the earlier stages the range of possible options is very broad. It is important to recognise that one option to achieve the objective may be to carry out what is, in engineering terms, a different project. Any transition must be managed to ensure that the changes to the engineering are reflected in the estimates, programmes and business case.

As a project progresses through its feasibility stage there will be one or two project options, usually one, which proceeds to a detailed design. It can be seen therefore that the nature of the risks change from broad-brush issues such as the type, size and location of the project to a narrower range of issues. When one of them is selected the emphasis changes to the much more narrowly focused on the estimation of realistic cost forecasts, and the detailed design and the preparation of a detailed programme for the execution of the project to achieve the best value for money. During the implementation phase the range of risks narrows still further to those associated with the procurement, manufacture and delivery of materials and site construction activities. Management of each element of the project and its associated risks may be facilitated by adopting specialised techniques.

Broadly speaking, the earliest phases of the project are concerned with value management to improve the definition of design objectives; the design stage is concerned more with value engineering to achieve necessary function at minimum cost; and the construction phase is centred around quality management to ensure that the design is constructed correctly without the need for costly rework. Systems engineering may be used to manage the technical issues and interfaces on such complex projects.

It is important to realise that each phase will contain a number of key assumptions, which are made to allow the project to continue. As the project progresses firm information will be available to replace these assumptions. Sometimes this information will be different from the original assumption, which it supplants. It is important then to reassess the project and see if this changes fundamentally the basis for the previous work and also what impact this could have upon the future development of the project. From time to time completely new risks may arise. However, risks should diminish as the project progresses. It is necessary to ensure that risks, which have not occurred and can no longer occur, are removed from future assessments and analyses and are also removed from registers and reports, to assist in managing risks.

One further point, which is a major risk for many projects must be made. It is that in reality, projects are not always continuous. There are breaks and discontinuities in its life cycle. Frequently this is because funds are not available to finance the next phase, the market changes or other circumstances change. The last two can occur even if there are no discontinuities in the project, which is why periodic reviews of the project are essential.

Projects, which are known as *fast-track projects*, compress the normal project phases and overlap them to some extent. This is true of oil and gas projects where the detailed design of piping and equipment may continue some time after the design of the civil engineering works such as foundations and structures. This approach has been adopted increasingly by the building sector and to a certain extent by the civil engineering sector – in the design and build approach. DBFO and PPP projects take this approach even further by shifting the assessment of business risk as well as that of detailed design and construction risks to the front end of the project. While the totality of this approach is perceived as being beneficial in shifting the bulk of the risk onto the private sector contractors, it has the disbenefit of reducing the step-by-step approach to management and decision making which flows from the traditional multiphase approach.

In some cases however, such as the Channel Tunnel rail link, a hybrid approach can be adopted. In this particular case, the difficulties of routing the track through Kent, and in and around London, meant that the selection of the route was a highly critical task. Therefore, all the front-end work was carried out by the owner. The private sector companies who tendered for the detail design, a construction and operation of a service, had been given an outline design together with an estimate and a timescale. Their task was first to examine the information, prepare a business case including the consideration of raising additional revenue through improving the existing Eurostar international train service, property development and running additional services into London from new stations in the suburbs. The review of this estimate and programme together with the risk assessment is described as a case study later in the book.

2.6 Project appraisal

From the viewpoint of risk management, the appraisal phase is the most crucial. It is during this phase that the key decision regarding the choice of option is made, although occasionally there may be more than one option selected for more detailed review.

Appraisal is the process of defining first the alternative ways of achieving the project's objectives, that is to say, defining the options that are available and choosing between them.

A fundamental prerequisite is that the project's objectives have been set. Ideally, one will dominate and its influence on the other objectives must be clear. All the members of the project's organisation as well as all the stakeholders must be in agreement on this point. It is a fundamental risk that if the objectives are not clear, not agreed or not communicated to those involved, the chance of the project being a success is reduced because the potential for changes and conflict is increased. In such cases value management is useful to develop and clarify objectives.

The public and private sectors may have differing views on objectives and hence on viability. For example, the private sector may consider that early completion and entry into a competitive market should be the top priority. This implies that risks related to the project's timescale and programmes are key. In other cases, the final performance and/or quality of the project may be paramount. But most often, cost and affordability, that is to say the amount of finance available to fund the project, will be the dominant factor.

However, it must be clearly understood that there is a trade-off between these parameters. For example, for a given level of performance there is likely to be a narrow range of project durations, which are commensurate with minimum cost. If the project is required earlier, the cost is likely to be higher because the project is effectively being accelerated. If the period is longer, perhaps because funding limits the resources that can be devoted to the project, then ultimately the cost is likely to be higher because the time-related cost of those resources and management effort increase. In periods of high inflation, either general or industry specific, the effect of delay is multiplied. It is necessary to study and predict trends in the market and the economy, anticipate technological developments and the actions of competitors because these are areas of significant uncertainty and hence risk. This may be called *market intelligence* related to the commercial environment in which the project will be developed and later operated. The impact of changing costs and timescales on the business case must be taken into consideration because delayed completion defers income and benefits.

Other major considerations during project appraisal are:

❑ The estimates of cost, both capital and operating. Single figure estimates are inadequate to represent the range of possible outcomes, due to general uncertainties and specific risks.
❑ The project execution plan which should give guidance on the most effective way to implement the project and to achieve the project objectives, taking account of all constraints and risks. This plan should define the contract strategy and include a programme showing the timing of key decisions and award of contracts.

It is widely held that the success of the venture is greatly dependant on the effort expended during the appraisal preceding sanction. There is, however, conflict between the desire to gain more information and thereby reduce uncertainty, the need to minimise the period of investment and capital *lock-up* and the knowledge that expenditure on appraisal will have to be written off, if the project is not sanctioned.

Expenditure on the appraisal of major engineering projects rarely exceeds 10% of the capital cost of the project. The appraisal, as defined in the concept and brief accepted at sanction will however freeze 80% of the cost. Although often sought, the opportunity to reduce cost during the subsequent implementation phase is relatively small.

Appraisal is likely to be a cyclic process repeated as new ideas are developed, additional information received and uncertainty reduced, until the promoter is able to make the critical decision to sanction implementation of the project and commit the investment in anticipation of the predicted return. It is important to realise that if the results of the appraisal are unfavourable, this is the time to defer further work or abandon the project. The consequences of inadequate or unrealistic appraisal can be expensive or disastrous. This may be because the appraisal identifies risks that are likely to have significant impacts during the project's implementation phases or during its operational phases. If this is the case and the project appears not to be viable, it must be thoroughly reviewed. A greater risk may imply a higher return on the investment. Whatever the result of the appraisal the decision on how to proceed should be based on its findings, even if this means abandonment. This decision cannot be shirked. The project should not be sacrosanct if, on a rational analysis, it is unlikely to succeed or represents too great a risk.

2.7 Summary

It must be noted that the application of risk management techniques is likely to result in an increase in the project's capital cost and implementation programme. This is because estimates and plans prepared during the pre-feasibility phase of projects are likely to be low, because little detail exists and it is human nature to be optimistic at the start of any new enterprise. Recently the term optimism bias has been coined to cover the overrun caused by overoptimism or systematic failure to expose risks.

Ideally, all alternative concepts and ways of achieving the project objectives should be considered. The resulting proposal prepared for sanction must define the major parameters of the project – the location, the technology to be used, the size and type of the facility, the methods for operation and maintenance, the sources of finance and raw materials together with forecasts of the market and the predictions of the cost/benefit of the investment. There is usually an alternative way to utilise resources, especially money and this is capable of being quantified, however roughly.

Investment decisions may be constrained by non-monetary factors such as:

- ❑ organisational policy, strategy and objectives;
- ❑ availability of resources such as manpower, management or technology.

The key to realistic appraisals is to create a level and unbiased basis for all the options for the purposes of comparison between them. Therefore the estimating techniques, programmes and assumptions used for each option should be the same, based as far as possible on closely similar base dates for costs. This facilitates direct comparison between the options.

The impact of risks related to particular options can then be assessed to provide a full comparison. It is the impact of these risks, which may differentiate between the options. For comparative purposes, risks that impact equally on all the option may be ignored unless they are critical in assessing the viability of the project. In reality, the assumptions that are made about risks and their avoidance or mitigation through risk management can be just as biased as unrisked estimates. The outcome of any appraisal should be treated as indicative, not absolute. The future cannot be predicted accurately, regardless of the sophistication of predictive method and tools.

Chapter 3
Understanding the Human Aspects

Risk management is a proactive approach to the *what ifs* that can determine and influence the project's outcome and achievement of its objectives. It is well known that unforeseen events called risks will happen during the lifetime of a project and some of these can seriously damage the project. Risk management is about avoiding, reducing, absorbing or transferring risk and exploiting potential opportunities.

Projects evolve in rapidly changing environments because of the pace of technological development, increasing complexity, new methods and tools, new markets, increased competition, novel business opportunities and demanding customers. This implies that construction projects are dominated by objectives based on time, cost and quality; as discussed in Chapter 1. Good project management has to some extent always been concerned with project uncertainty when establishing cost estimates and schedules, it is clear that there have been shortcomings in the approaches adopted, and in future projects will require much more systematic and effective risk management.

It is true that there can be problems in describing project risks to management, and convincing stakeholders that money should be spent on the day to avoid risks that can happen in the future. This underlines that the success of risk management in practice is about human and organisational factors such as understanding, motivation, attitude, culture and experience. The increasing awareness of the need to manage risk effectively along with its awareness in corporate governance has eased the introduction of risk management.

The quality of project risk management relies on a number of factors, including management attention, motivation and insight among project personnel, the qualifications and knowledge within the project and the experience and personality of the manager and/or risk analyst(s) leading the process. These four key success factors are directly related either to people or to how the project organisation works. Again, one of the keys to success in risk management is to understand people and their behaviour in different roles.

3.1 Risk management – people

People play many roles and their behaviour changes with the role played, for example from work to home. This implies that an individual's attitude changes with different external requirements, constraints and expectations.

At home as well as at work most of our everyday choices are affected by risk, and we constantly perform some kind of risk management when we make decisions. Crossing busy streets, getting married and buying a house are all personal risks most people face. It is sometimes thought that by not taking an action risks can be avoided, but this means that by so doing an opportunity might be missed. Many people also seek risks because the uncertainty about an outcome of an activity can provide excitement. People engage in risky recreational activities such as bungee jumping, parachuting and skiing to the South Pole. People play the stock market and they gamble, partly because of the stimulation that accompanies the risk and partly because of the chance of winning.

Risk is very much related to personal attitudes. There are two main categories of people, the risk takers and the risk avoiders. In general terms, entrepreneurs and investors are risk lovers, while people who take on a low paid but safe job and people who invest all their money in savings accounts are risk averse. A risk taker would accept a higher exposure and therefore a higher variability in payoffs. There are also differences in the way risk takers and the risk averse perceive risk. Risk takers tend to underrate risk, while the risk averse see all the *obstacles* and tend to overrate risk. Fortunately, we find people from both categories managing projects and working in project teams. This is the starting point for establishing a realistic risk picture of a project, and achieving a proactive management attitude toward project risk and its implications.

In companies, the project management methodology established does not readily accommodate the increasing requirements for risk management and many projects are therefore not set up to manage risk. Many companies believe that project management is about focusing on time and cost as fixed goals. A successful project finishes on schedule and at the budgeted cost, and performance is measured against objectives that are established well ahead of the project execution phase but often based on scarce information. In addition, some of these organisations think they do not have the time for risk management, but instead they spend vast amounts of time and money on correcting projects that deviate from rigorous plans.

Managing risk does not usually seem to be a problem for people on a personal level, although in some circumstances it can challenge senior management's perceptions. It seems much more a problem or a challenge to perform risk management within the organisational and methodological constraints of business and industry.

3.2 Risk management – organisations

Adopting risk management as part of the management philosophy depends very much on the people responsible for maintaining, performing and developing management guidelines and procedures in a company that is the managers themselves. Because of this, many companies benefit from having innovative managers who encourage risk management, but many suffer from management, which is averse to it. However, failure to undertake risk management in an explicit and formal manner as a routine aspect of project management is increasingly regarded as commercially unacceptable.

Many project managers perform risk analysis because somebody else, for example their client, the parent company or the government, has told them to do so. The work is performed in a hurry and used as an alibi, in case things should start going wrong. This is a very common approach, and again it is a management failure; they do not understand why they should perform risk management and the benefits that can be obtained.

The need to perform a risk analysis can emerge from planners and cost estimators, but the project manager must understand the benefits that can be gained and that resources must be used in the risk process. The project manager or the person responsible must help create the environment for the analysis, such as underlining the importance of performing the analysis, express goals and expectations, actively be part of the risk process and be responsible for the actions or responses resulting from the risk assessment and analysis. The project manager should make sure that all key personnel are available and have included their input to managing risk in their schedules.

One of the main obstacles when introducing risk management to an organisation is the lack of openness and communication within the organisation. Performing risk management in such organisations can be very painful in the start of the process, but the risk process has proven to be a catalyst in breaking down communication barriers and to provide an environment for openness and discussions.

Successful risk management within organisations relies on management attention, motivation, a methodical approach, project management methods, competence, knowledge and understanding, culture and openness. It is not, and cannot be, a substitute for project management itself. However, the art of performing risk management is not learned on a three-day course or by handing the responsibility for managing risk to a person without risk experience. The organisation should also understand that managing risk is a matter of learning and improving over time. The organisation should recognise itself as a *learning organisation* on the risk issue, and hire consultants to perform the first risk analysis and to train the personnel until maturity in managing risk is achieved.

3.3 The risk management process

The risk management process focuses on the needs and the priorities of the client and includes methods, techniques and tools especially developed for this purpose. The process is often headed by a risk manager or analyst who is responsible for establishing a framework for extracting information from project key personnel through risk identification and assessment.

The key to success in the process is the contribution from the people working in the organisation. Risks are most commonly identified and structured in open-minded creative workshops facilitated by the risk analyst. Based on the data collected and available project documentation, response plans (treatment plans or action plans) can be developed. To gain understanding of the project level risks and develop realistic baselines for the schedule, cost estimates and contingency provision, a risk model is created, most commonly with the aid of a risk analysis software tool. Input and results are verified by the project team and, if necessary, by external resources. The process is iterative with loops back to previous stages that secures verification and project team ownership. The Association for Project Management's Project Risk Analysis and Management (APMPRAM) guide gives a detailed explanation of the process.

Risk management relies on a formal process for identifying and quantifying the subjective judgements of experts and project personnel. The risk analyst facilitates drawing risk information from the participants, creates an analysis showing the effects of risks and presents the results back to the participants. They must agree with and own the output from the assessment or risk analysis. If they reject the results, they will not be willing to work with the results. Ownership of results is therefore vital no matter how sophisticated the software is. Commitment and ownership can only be gained through close co-operation and a good relationship between the analyst and the project team.

3.4 Some guidelines to the risk management process

The most common way to perform a risk assessment is to gather key personnel for risk identification sessions, and then interview them in groups or as individuals. However, both the risk analyst and the individuals will bring bias to the results. This bias should be minimised by ensuring that sufficient people are involved in the process. A key rule is that groups make better decisions than individuals, and in addition, groups create stronger ownership to risk assessments and the results from analyses, although groups can be less risk averse than individuals and can be dominated by an individual or a small number of participants.

One of the most important factors in the risk management process is the gathering of key personnel with one purpose only; to discuss, assess and if possible quantify the risks that may affect the project's objectives. Such a group process stimulates participants to communicate and express their opinions in an open-minded environment where people are free to express whatever feelings they have. The group will most commonly include experts from various disciplines who can contribute to the risk assessment, which should lead to fruitful discussions and communication across the project organisation. The process should be headed by an experienced risk workshop facilitator to make sure that the necessary information is collected. The workshop should be complemented with interviews of key personnel to try to avoid or understand any biases in the group that may influence the results.

Group or individuals

A group can be defined as 'two or more individuals who are interacting with one another in such a manner that each person influences and is influenced by every other person'. An important question is, whether individuals working together in a group perform more successfully or efficiently than individuals alone. The answer to this is complex and depends on the task, the structuring of the task and on the organisation responsible for making judgements about the task.

The judgement of a highly skilled individual will often be more accurate than the combined group judgement. However, in general, group judgements are seldom less accurate than the average individual judgements and are often superior. In addition, the group often has the advantage of a wider range of knowledge and should therefore generally yield judgements that are more realistic.

Social psychologists have put some effort into specifying the *ideal size* of problem solving groups and concludes 'groups of five are the most effective for dealing with mental tasks, in which group members collect and exchange information and make a decision based on the evaluation of this information'. This is no rule, and in practice the group size will depend upon the complexity of the task and the availability of key personnel. Some tasks may require only one person, and groups can also be composed of more than five persons when the task is structured.

The group approach is used to avoid well known pitfalls such as; motivational bias (for example individual estimates reflect the wishes of management) and non-representative experiences. If you transfer general knowledge and experiences to a specific situation without adjusting to the differing circumstances and risk, the use of experience and knowledge from your last project will dominate over experiences gained on previous projects. This is likely to result in a lack of imagination, which could mean that all upside potential and downside risk is not foreseen because information is neglected – for example, the factors that went very wrong on the previous project because the current project will be different and those factors will not occur again.

As noted above, groups do have shortcomings. They tend to converge on a judgement and counter arguments are regarded as hostile. *Groupthink* is simply a tendency to seek concurrence. Four conditions are likely to foster groupthink:

(1) groups that are isolated from the judgements of qualified outsiders;
(2) groups with a strong leader and the procedures for debate are not established;
(3) a lack of a methodical approach;
(4) where there are immediate pressures to reach a solution.

Groupthink can be prevented by encouraging dissidence, call on each member of the group to be critical and reinforce members who voice criticism of a favoured plan.

A group discussion often produces a shift in individual opinions. Such a shift is not necessarily in the direction of greater risk. If the initial opinion of the group tend towards conservatism, then the shift resulting from group discussion will also tend towards an extreme conservative opinion. In such a case the group will usually shift towards the pole favoured by the majority. Motives for this behaviour include the desire for a favourable evaluation by others and a concern for self-presentation. This effect may be particularly evident when members of the group expect future interactions with one another in other settings.

3.5 The risk workshop

This section describes an approach to facilitating a group workshop.

Preparation

Imagine that you are the risk analyst, entering the room to meet a group for the first time. The group has been told that they are part of a risk analysis but have little or no experience from previous risk analyses. The group consists of five people with different backgrounds, experience and expectations. You can see on their faces that they are sceptical, and if you start the identification phase right away the project will seem to be a very safe one. This is often the situation, which risk analysis consultants and/or facilitators face.

The first thing to do is to set the group thinking about risk and uncertainty. Start by showing the group some examples of successful projects and failures. Measuring in cost is a very efficient way to underline the importance of undertaking risk assessment and analyses. Try to describe the performance of some recent and local projects that are well known to the participants, such as these in Table 3.1.

The percentage figures in brackets describe approximate changes from the initial estimates. Do not put much trust in the figures. When you prepare this exercise, you will find it very hard to find successful projects, while it is very easy to come up with poor projects.

Warm up exercise 1

Ask the group to identify five projects that have been successful and five that went wrong. They should also list the main reasons for either success or failure. The group will most likely come up with a lot of project failures and some successful projects. This exercise will shift their minds

Table 3.1 Relative performance of recent projects.

Failures (cost increase)	Success (increased profitability)
Sydney Opera House (7000%)	Microsoft Windows (a lot)
Concorde (1500%)	Hong Kong toll road (1000%)
Channel Tunnel (1000%)	Oil and gas projects (10–30%)

Table 3.2   Exercise 2 estimation table.

	Prices	
Items	Today	Year 2020
Tabloid newspaper		
Levi 501 jeans		
Three room apartment or house		

towards risk thinking of projects owned by other companies. It is a lot easier to see what others have done right and wrong.

Warm up exercise 2

Take their minds off the project risk analysis by inviting them to join in another little game. This exercise is about coming up with prices on items we buy on a private level. Ask each participant to fill in Table 3.2.

You are most likely to get different prices from most participants, that is, you get a range of outcomes for each item (a distribution). For the Levis jeans and the apartment, the spread can be very high. This is a very good and simple lesson to prove that risk affects most of the things we do, that the assumptions on which the prices are based are very important and that it is important to focus on the critical risk. In this case, if you are buying all three items, focus on spending your time reducing and exploiting the risk affecting the house or apartment investment!

The group should at this stage be at ease and have their minds open and ready to be innovative in the risk identification phase of the workshop.

Risk identification

It is now important to draw parallels between the project, which is being assessed, risk and uncertainty and exercises 1 and 2. Start the real identification stage by making all participants write down what they think are the main risks in the project. Give them approximately 30 min to complete this exercise.

When they are done, you should write all of the identified risks on a whiteboard or flipchart. Then go through each risk with the group to create a common understanding of the importance and magnitude of each risk. Categorise them, for example, using the following three categories shown in Figure 3.1.

An experienced facilitator should be able to apply their own checklists to go through risk factors that the group has not yet thought about. When

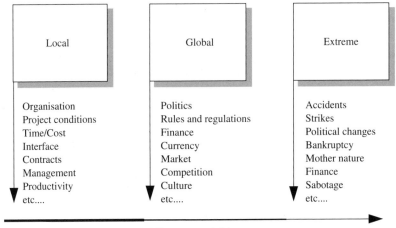

Figure 3.1 Hierarchy of identified risks

this session is complete, the group has probably identified 80% of the risks. The last 20% may not be worth any further effort, and a few of these risks could not reasonably have been foreseen.

Quantification

The risk management process now enters into the quantification stage. The risk quantification process is very much based on subjective judgements from project personnel. Empirical data are usually insufficient to quantify the uncertainty in the consequences of a course of action, and judgmental probabilities provide a logical means for overcoming this limit. Making judgements of uncertainty quantitatively minimises ambiguity: the statement 'the activity will most likely finish after 30 days, with an optimistic duration of 20 days and a pessimistic duration of 45 days' is much less ambiguous than 'we think there may be a delay'. Another way to approach this is to recognise that in deriving a component of an estimate, the estimator will consider a low value and a high value, finally choosing a figure between the two, that is to say, that 66% of the available data is discarded. It is useful to agree the metrics for the assessment, for example, what cost impact, schedule impact and the like are categorised as high, medium or low.

It should be noted that in the early stages of projects, the estimate and schedule are unlikely to have been developed in detail, so risks cannot be quantified with precision. There is a real temptation to develop sophisticated and detailed risk models that are not commensurate with the level of understanding of the project. The quality of the output can never be better than that of the input.

A risk model is usually created in a risk analysis software tool. The quantitative risk is most commonly included in the risk model by estimating a pessimistic, a normal and an optimistic value together with a probability distribution. Do not expect that participants in the process will have strong feelings of which probability distribution to select. Therefore, you as the analyst should select simple distributions, for example, the triangular distribution, although care must be taken to oversimplify. Incorrect choices of input data and distribution may significantly reduce or negate, the value of the analysis.

It is very important that the risk analyst manages to transfer the information gathered in the identification phase into risk assessments reflecting the real risk affecting the parameters. Often the spread is far too conservative, that is, the risk is underestimated. Often the high value does not include the worst case or anything approaching it, and the low value does not include opportunity for improvement (assuming the base figure is not so optimistic that it cannot possibly be improved on). This can be overcome by separating the base range of uncertainty (estimating uncertainty) from the significant discrete risks so that these can be separately assessed and modelled. It is also very important to discuss the assumptions behind the estimates to ensure that the risk assessments are anchored to the estimates. However, it is frequently difficult to separate allowances (contingencies) within the base estimates from additional provisions that must be made.

There is a lot of literature describing how to quantify risk, see also references. It is very important to use a practical and approximate approach when quantifying risk and selecting probability distributions. It adds little value to turn the project into a complex mathematical equation.

In fact, the NAOs report 'London underground PPP: were they good deals?' noted that:

> 'With hindsight London Underground agrees that some of the cost [of the Public Sector Competitors] particularly the production of refined cost projections, extensive Monte-Carlo simulation and overly detailed documentation associated with the model's development was unnecessary, given inherent weaknesses in the underlying data.'

3.6 Communication

The next *human* phase in the risk management process is the communication and understanding of the results achieved from analysing the risk model. The technical aspects of modelling and simulation techniques are described elsewhere in this book.

The results achieved in the *analysis* phase should now be communicated to the project managers and the project key personnel involved in the analysis in order to validate the results, and to clearly see the effect of the risk affecting the project goal.

The results should be easy to understand. Start with presenting the main risk assumptions and the risk assessments. Focus on the important results and do not use difficult statistical parameters if you are not confident that all members of the project team understand these. Some of the results will be surprising, but remember, results are not debatable as long as the project team agrees with the risk assumptions, the model and the risk assessments. Risk analysis is a process, which is very often iterative and therefore some adjustments to the initial risk assessment must be made. It is very important when presenting results to go back to the main initial risk assessments to clarify that these still reflect the project risk. If not, they should be adjusted and new results produced.

It is essential to differentiate between assessments and quantifications of risks that are the unmitigated impacts, and the mitigated impacts following implementation of actions to avoid or reduce the risks. Modelling can be used to demonstrate the effectiveness of mitigations.

3.7 Summary

A risk assessment or analysis is only successful if actions are initiated based on the results. It is therefore very important that the risk management process be handled in such a way that the project team feel that it *owns* the results and that it is their responsibility to actively manage the risks. They need a high degree of involvement in the identification and assessment phases, and must accept and own the results emerging from the analysis, the results of which must be clearly understood and communicated within the project organisation. This is crucial for projects that want to successfully introduce risk management to improve the likelihood of successful delivery.

The key to achieve a proactive risk management attitude within a company project or a programme relies first on the people involved. The success factors are management support, motivation, insight, openness, involvement of key personnel and learning. These factors should be combined with a risk management method that focuses on participation, ownership and responsibility. Risk management must not become just another bureaucratic task that takes time but add little to delivering the project. The risk process should focus on the identification and response phases, and not on the creation of advanced mathematical models of the project. The risk management initiative must come from management which really understands what risk management means to the company, and the method should be introduced as a competitive edge and as an integral part of the management philosophy.

Chapter 4
Risk and Value Management

4.1 Introduction

Construction projects should only be sanctioned following a careful review of need. Many projects suffer from poor definition and inadequate risk analysis. Value management is primarily about enhancing value, which often requires a thorough understanding of the relationship between cost, risk and the associated benefits or profits derived. Linking both risk and value management to form an integrated management system, needs to be considered during the project inception and appraisal stage. Instigating risk and value management at this stage leads to a greater likelihood of risk and value management practices throughout the remainder of the project promoting commercial success.

Central to the relationship and management of risk and value is the concept of value for money (VFM), and its assessment, which is carried out to determine efficient contract strategies and project solutions. As projects become more holistic in nature as per sound value management philosophy, pressure is placed upon the assessments made about the quantum of risk included in specific contract strategies and project solutions.

This chapter illustrates how project stakeholders – those being the investors, designers, contractors, operators, end-users and others, with an interest, and power to influence, the project outcome – may use value management to determine the type of contract strategy most suitable for a particular project by identifying and allocating risks associated with meeting the project's objectives. An integrated management system for the appraisal of risk and value in a project is outlined, drawing upon specific risk and value management techniques.

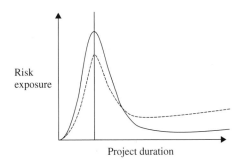

Risk
exposure

Project duration

Figure 4.1 A comparison of two project solutions in terms of risk exposure.

4.2 Approaches to the management of risk

At the start of any project, there is a large amount of risk resulting from the uncertainty surrounding the way in which the project will proceed. However, despite this large amount of risk, the real exposure to these risks is minimal since little has actually been committed. Once the project moves from the feasibility to the implementation phase, the stakeholders are then exposed to these risks because decisions have been made and work has started on the project. This level of risk exposure then decreases as the project proceeds since the amount of risk remaining in the project decreases as the project proceeds. The reasoning behind this is that the risks are either realised or do not occur. However, projects with long life-cycles may find that the risk exposure begins to creep up, as refurbishment or maintenance work threatens the operational performance of the asset. This difference in asset performance is illustrated in Figure 4.1.

The level of risk is a combination of the probability of occurrence of the risks and their possible impact on the project should they be realised. This assessment need only be qualitative and subjective, but the information that it provides serves as a guide to which risks require further investigation or analysis.

There are two basic types of approach to the management of risk in projects, and these are the informal approach and the formal approach. The type of approach adopted influences the procedures and processes that will be used in the management of risks in typical engineering projects.

Informal approach to the management of risk

The informal approach to the management of risk is one that views the risks in a subjective manner and due to the nature of this approach, many organisations implement these methods but do not realise that they are operating any kind of risk management procedure. The main danger is that this approach is deemed sufficient and experience shows that it is not.

One of the most widely used techniques in the informal approach to the management of risk is the provision of contingency funds. There are two main types of contingency fund those being lump sum contingencies and percentage contingencies. A lump sum contingency is a sum of money put aside, in the project budget, in case any extra money is required during the project. A percentage contingency is similar to a lump sum contingency but, instead of being a fixed sum of money, it is a percentage of the total project cost, included in the project budget.

Contingency funds can be used as a risk management technique because the amount of money allocated to a contingency fund should be representative of the cost of mitigation of the risks that are thought likely to occur in a particular project. A contingency fund is not financially representative of all the identified risks in a project because it is unlikely that all of the possible risks would be realised. Hence, a contingency fund should represent the cost of the risks that are thought likely to be realised in a project rather than, as is often the case, being adequate to cover all eventualities. Often, to compound this misunderstanding, contingencies are thought not be needed, that is to say the contingency is available, but is not to be spent.

Other informal procedures for the management of risk involve discussions with experts or people with experience on similar projects and assessing their views as to the possible risks in a project, then reviewing the project in the light of these possible risks.

Formal approaches to the management of risk

The formal approach often consists of a set of procedures laid down by an organisation for use in the risk management process. These procedures are structured and provide guidelines to be followed, so that they can be used by any member of the organisation. This enables a uniformity of approach to be achieved. This formalising of the risk management procedures ensures that the process is more objective than the informal approach. Most authors recognise objectivity as an essential feature in the process of managing risks.

Formalised procedures for the management of risk in projects are designed to suit the needs of the particular organisation; hence there is no single methodology. However, there are frameworks for formalised risk management procedures, which do not detail the methods that should be used, but allow the user scope for choosing appropriate techniques. The quality of a formal process of risk management is generally accepted to be dependent upon management awareness, motivation among project personnel, a methodical approach, the information available (often linked to the project phase), the assumptions and limitations upon which the risk analysis is based, the qualifications and knowledge within the project and the experience and personality of the risk analyst(s) leading the process. Similarly, there are a number of well-known assessment pitfalls including: (1) management bias that occurs when an uncertain variable is viewed as a goal rather than as an uncertainty and; (2) expert bias where experts are expected not to be uncertain, but to be sure of things. This may lead to underestimation of uncertainty.

Qualitative risk assessment

Construction projects usually involve a large number of activities and events, which usually contain uncertainties due to lack of resources or data. Projects should only be commissioned following a careful analysis of the need and risks perceived. Failure to think through the needs and risks associated with a project may cause problems through the adoption of a contract strategy not best suited to meet the needs of the project and provide an equitable allocation of the risks identified.

A typical qualitative risk assessment usually includes the following issues:

- a brief description of the risk;
- the stages of the project when it may occur;
- the elements of the project that could be affected;
- the factors that influence it to occur;
- the relationship with other risks;
- the likelihood of it occurring;
- how it could affect the project.

Quantitative risk assessment

The probability of a risk arising is a key factor in the decision making process. Possible consequences of risk occurring are defined and quantified in terms of:

- *increased cost* that is additional cost above the estimate of the final cost of the project;
- *increased time* that is additional time beyond the completion date of the project through delays in construction;

❑ *reduced quality and performance* that is the extent to which the project would fail to meet the user performance based on quality, standards and specification.

These may be analysed using sensitivity and probability analysis.

4.3 The standard risk management model

There are many models or methodologies for managing the risks in projects. Most risk management practitioners have developed their own model or methodology that is best suited to the types of projects that they are involved in. Nevertheless the methods commonly used in the United Kingdom are based on HM Treasury (2003, 2004a and b) guidance documents, which provides a framework that contains the steps to be taken in the project risk management process. However, the choice of techniques to be used is for the user to decide. The model is essentially designed for use in the construction industry; however, it is suitable for use in most industries provided that the techniques used are chosen specifically for the project and the industry in which it is to be used.

The standard model is divided into four parts: (1) risk identification, (2) risk analysis, (3) risk response and (4) risk review. The sub-processes and their control are shown in Figure 4.2.

The value and risk management project appraisal model presented focuses heavily upon the identification, analysis and response to risk. However, the process of risk review is essential to maintain and improve future appraisals and assessments of projects. It also influences the value management proceeds in this case-option appraisal, by allowing the users to consider specific options used in the past on similar projects, making

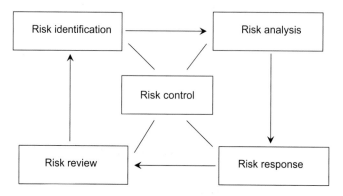

Figure 4.2 The risk control process (Merna and Lamb, 2004).

them aware of their weakness, and strengths, and shortening the time taken to develop viable solutions based on the risks facing a project. Additionally it identifies specific risk allocation structures in association to contract strategies, providing more depth in the assessment process.

The approach proposed is very flexible because it gives the user the freedom to choose techniques that are appropriate for a particular project industry based on the level of detail available.

4.4 Applying risk and value management

There is no single correct approach to the application of value management techniques especially when combining them with risk management; Merna and Lamb (2004). Although, most projects and their procurement strategies vary, there are a number of stages common to projects themselves as demonstrated in Chapter 2. Some of the stages overlap depending on the type of project and the method of procurement; typical stages are:

❏ definition of objectives;
❏ understanding the project;
❏ applying value and risk management;
❏ VFM and iteration of the process.

The implementation of the last stage is dependent upon resources and time available for the project appraisal; to illustrate how this process may be completed during the appraisal of a project the following process diagram, Figure 4.3, is presented.

Many organisations utilise value management in conjunction with project evaluation as a means of achieving value for money for the project stakeholders. To enable the representative models to facilitate influence over the project such value and risk should be identified as early as possible in the process as per figure. The values held by the project often fall into three categories those being:

❏ internal values (important to the project owners);
❏ intermediary values (important to the project delivery);
❏ external values (important to the customers).

Risk may fall into a similar structure of categories, allowing risk to be a transferred and allocated to those with alternative perspective on

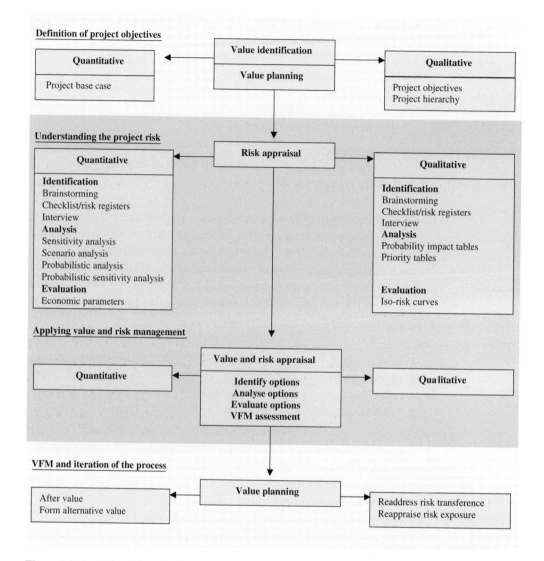

Figure 4.3 Applying risk and value management in the appraisal of a project.

the valuation of risk. The categorisation of risk in this fashion also operates with regards to the type of contract strategy considered.

Therefore, value management in terms of contract strategy selection is the identification of a contract strategy that safely minimises the valuation of risk to the contract without jeopardising the integrity of negotiable terms for the distribution of internal, intermediary and external risks throughout the project's lifecycle. Furthermore, as the valuation of risk is highly dependent upon the allocated participants' interpretation of risk, inappropriate or inexperienced valuations of risk must be identified and addressed as soon as possible.

4.5 Value management processes

Value identification

Brainstorming, interviews, prompt and checklists may all be used to identify the objectives that are critical to the completion of the project. This may be stored in a value register, listing and prioritising those objectives identified. In some cases, an objective hierarchy is prepared, which ranks objectives in order or on priority. From this objective hierarchy a number of risks can be identified against each objective.

Value planning

The value plan establishes how the objectives address the future appraisal of the contract strategy or project proposal, outlining the quantitative and qualitative systems to be used. This plan provides the basis for an audit and accountability trail throughout the development of the project identifying where, why and how specific decisions were made about the objectives and risks inherent in the project. The plan should also cater for changes made to either the priority or objectives listed for the project. The responsibilities of the stakeholders are identified along with the areas of potential conflict and key project constraints.

This first review should therefore result in:

- confirmation that the project is required;
- identification of the objectives and priorities of the project;
- a favoured option for the further development of the base case model.

VFM may be represented both qualitatively and quantitatively. This places pressure on the interpretation of the results, thus requiring the value plan to place in to context the metrics used by both qualitative and quantitative techniques for the assessment of VFM. Thus, the value plan must also identify those values (objectives) that may be represented in a quantitative and qualitative manner and the associated risks (negative and positive) ability to be represented in a quantitative or qualitative manner.

The project stakeholders can now appraise the risks based on the base case cost estimation model.

4.6 Understanding the project risk

Initially, this has been simplified as only the base case model is considered for analysis. The base case model is a proposed solution that meets the minimal requirements set against the objectives defined from which an exposure to risk may be gauged.

Risk identification

Brainstorming sessions involve getting the key project stakeholders together to identify and prioritise the risks in the project. This technique enables the stakeholders to hear what the other members of the project team see as risks and to use these ideas to inspire them in identifying additional project risks. It is important to choose carefully the people who are to constitute the brainstorming group, as there needs to be the right mix of project personnel with appropriate experience and seniority to ensure a successful session.

Interviewing project personnel from each discipline and staff within the organisation who have experience of similar projects, ensures that corporate knowledge and personal experience are utilised in the process of identifying risks. This technique allows project personnel to have their say about the risks that they can see in the project, and gives them a feeling of involvement in the process and ownership of the identified risks. This should then lead to a greater acceptance of any measures implemented to reduce the risks.

The examination of historic data from previous, similar projects helps to utilise corporate knowledge. However, an organisation may not have carried out a similar project, or the data from a previous similar project may not have been recorded; so this technique can only be successful in a limited number of cases. Database systems that actively manage and report the progress of projects may be a useful source of information. However, such systems are often limited in terms of the useable or relevant data being stored.

Risk registers are documents, spreadsheets or database systems that list the risks in a defined project and their associated attributes (positive and negative). The risk register also identifies defined events assigning a value to these events, which are dependent upon a probability of occurrence. They are flexible in terms of the system and information that may be stored in them.

Each project has many risks, which depend upon technology, country, organisation and institutional involvement and the contract and finance strategy applied, but the key sources of risks in projects are essentially

Financial risks
Legal risks
Political risks
Social risks
Environmental risks
Communications risks
Geographical risks
Geotechnical risks
Construction risks
Technological risks
Demand/product risks
Completion risk
Commissioning risk
Supply risk
Force majeure risk

Figure 4.4 Checklist of construction risk drivers.

the same. However, what does change is the involvement of specific stake-holders who stipulate specific management policy or practices before a project commences. This can hold significant implications to the overall success of the project.

A number of authors have listed sources of risks associated with engineering projects during the identification process. These sources of risks, the *risk-drivers*, could be used as a checklist, as shown in Figure 4.4.

These sources of risk relate to both project and non-project specific risks. Each of these sources of risk are generic and it is up to the individual or team to define the boundaries of these sources and then to breakdown these sources into detailed risk elements, so that there can be a common understanding amongst those attempting to identify the risks in a project. The division of risks into source elements can be difficult as the risks attributed to each source element are chosen by individuals and thus, this method is exposed to a large element of personal subjectivity. It can also lead to the possibility of *double-counting* some risks by attributing the same risk to more than one source; however, this may be beneficial in understanding the relationships between risk sources and elements.

Risk analysis

There are many techniques available for analysing risks; the term risk analysis has come to have different meanings for different people. In many cases, the perception of the term risk analysis has been shaped by the techniques to which they have been exposed. One explanation of the term analysis is the estimation of what will happen if an alternative course of action is selected. The process of analysing risks is important because it gives an understanding and awareness of the impact of risks on problems.

In this case, risk analysis is being applied to the base case model to establish the implications of risk on a standard model. This pilot operation of risk management prior to the application of value management helps to tune the future options desired, as it provides the team with an understanding of how risks are generated through the completion of the current objectives of the project. Therefore, the team becomes aware of solutions that would help either to reduce, mitigate or transfer the risk while safely protecting the value achieved from the project.

Due to the many different types of projects and the large spectrum of variables to analyse, there are wide ranges of risk analysis techniques available, it is important to choose the techniques appropriate for particular situations. Use of the same risk analysis technique for every project can be wasteful of time and money, by being too detailed for some situations and too superficial for others.

Each project requires a risk analysis technique that suits the needs of the stakeholders. There are a number of factors that should be considered when choosing the appropriate technique for a project or situation. The principal factors on which the choice of risk analysis technique should depend are the type and size of the project, the information available, the cost of the analysis and the time available to carry it out and the experience and expertise of the analysts.

Another factor in this choice is the purpose of the analysis. By carrying out a risk analysis, the possible effects of risks occurring can be seen in terms of the project's outcome. Prior to this, decisions must be made about the main priorities of the project. An example of the way in which this could happen would be, if the results of a risk analysis showed that, a project was likely to be delayed, but that if sufficient money were spent during the construction phase the chances of a delay would be dramatically reduced. This situation would then require a review to be made about the priorities of the project, whether it would be more important to finish the project within budget or on time.

Sensitivity analysis

The purpose of the sensitivity analysis technique is to answer the *what if* question by isolating the key variable(s) and evaluating the effects of incremental changes in the values assigned to the key variable(s). Sensitivity analysis is a quantitative technique, which allows the effect of economic changes in a project to be explored that is one of the best known non-probabilistic risk analysis techniques. A sensitivity analysis is carried out by identifying a project variable and giving that variable limits within which it is likely to vary. A number of points or steps, are examined around the deterministic value for the economic parameters. At each step, the values of the project economic parameters are calculated using the value of the variable at that step. This type of analysis can pinpoint the most critical areas of a project, in terms of the risks, and indicates where confidence in estimates is vital. The output from this analysis can be represented by a sensitivity plot or *spider diagram*, see Figure 4.5.

The *spider diagram* presents the information produced from a sensitivity analysis and clearly shows the variables to which the project is sensitive, assumed to be most and shows changes in the sensitivity of each variable. As an example, the outcome of a project may not be sensitive to a delay in the construction activities until the float in the project has been used, and after that point, the project outcome may then be very sensitive to any delay in the construction activities. This technique emphasises the point that variables can only be known within a certain range, which is defined by the person carrying out the analysis.

There are a number of limitations to the sensitivity analysis technique. The main limitation of this technique is that when changing a variable it assumes *ceteris paribus* (i.e. that all other things remaining the same), when in reality this is not likely to be the case. It assumes that only one variable changes at any one time and that there will be no corrective or preventative measures taken in response to any change in that variable. However, in reality if a variable was seen to be changing and affecting the project outcome then it is likely that some action would be taken to stop the change in that variable. A sensitivity analysis gives no indication of the likely range of change in the variable. The probability of occurrence associated with both the variable and the project outcome is not considered in a sensitivity analysis, although some practitioners have suggested the use of probability contours (see Figure 4.5), to provide more information about the risks. To carry out a sensitivity analysis requires the project to have been modelled, usually on a computer, prior to the analysis and so this technique requires the use of an experienced project modeller. The analysis does not take much time when using a computer, but the modelling of the project takes considerably longer than the actual analysis and this should not be overlooked.

Sensitivity analysis is very useful for identifying the variables to which the project is sensitive and those to which it is not initially sensitive but at a certain point does become sensitive to. A sensitivity analysis carried out in the initial stages of a project can provide useful information about where management attention should be focused during the project.

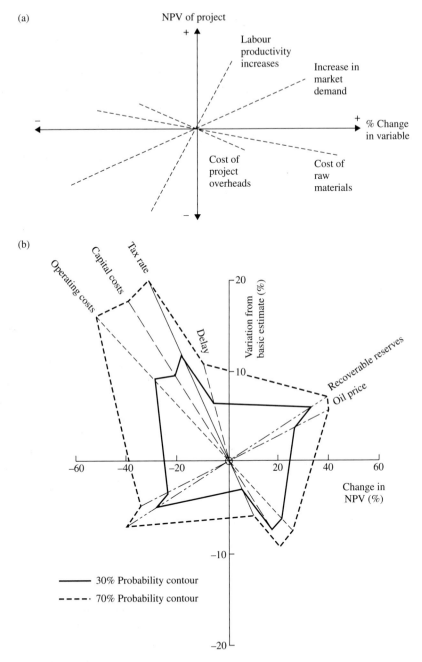

Figure 4.5 (a) Sensitivity diagram for major risks; (b) sensitivity diagram with probability contours.

This technique is especially useful for new and novel projects where the risks have not been previously analysed, and there are no previous projects to study. It is also useful to carry out a sensitivity analysis before carrying out a probability analysis, so that the possible effects of the variables on the project can be identified. This information can go some way to explaining the results produced by a probability analysis, particularly if the probability analysis has any unusual series of results.

Sensitivity analysis supports the value management process by focusing the attention of stakeholders on the variables that could jeopardise the value attained from the project. As depicted in Figure 4.3, this supports the value management process by analysing various project options that may all fulfil the project objectives. As some options may be more sensitive than others, the stakeholders can make an informed decision of the strongest option.

Scenario analysis

Scenario analysis embellishes upon a singular *what if* situation. The analysis takes into consideration a series of risk or variables, at the same time, addressing the weakness *ceteris paribus* of the sensitivity analysis. This analysis has been applied on several PFI/PPP projects and is consistently used for the development of business cases. In fact, NAO (2004) have identified the scenario analysis as a contributing factor to securing robust financial models, and several project refinancing and restructuring have increased the number of scenario analysis to support future financing structures. An example of a typical scenario analysis would be a 10% increase in financing costs, 10% decrease in revenue and a 5% increase in construction costs. This form of testing, establishes where stress may occur, as financiers place specific triggers often linked to financial ratios, to protect their interests in the project.

Probability analysis

Probability analysis assigns probability distributions to specific risks that may impact elements of a project's cash flow. Several iterations are completed based upon a series of allocated distributions to produce a frequency distribution of the expected outcome of the project. These are often reported in terms of economic parameters. Attaining sufficient information to support the assignment of distributions is difficult, especially under project conditions where information and its sources tend to be sporadic. This combined with correlated effects of risk can lead to a severe degradation of the usefulness of such models.

Probability sensitivity analysis

Probabilistic sensitivity analysis is a more complicated version of the sensitivity analysis technique outlined above, and involves assigning subjective probabilities to the alternative outcomes. This technique enables the user to see how sensitive the conclusion of the evaluation is to variations in the initially assigned probabilities.

To use this technique the analyst must know about the project in detail, probably gaining this knowledge from experience of previous similar projects. A lack of knowledge about the project when using this technique, could lead to deceptive results. In general, this technique is most useful for financial approval.

Probability impact

The collection of some information about the impact and probability of occurrence of risks could be undertaken using two simple matrices. In one matrix the user is asked to say whether they feel there is a high, medium or low probability of occurrence of each of the risks identified, and in the other to say whether they feel that the risk would have a high, medium or low impact on the project if it occurred. These matrices could be given to a number of people, particularly those involved in the project and those with experience of working on similar projects. The information given could then be distilled and plotted on a small grid, each risk being a point on the grid. This grid gives an immediate picture of the risks in the project and a qualitative assessment of their probability of occurrence and possible impact on the project. From this, it is possible to see the risks that are likely to have the least bearing on the project and those that require further investigation. Those in charge of the project have to decide which risks can be ignored and which need further investigation. In most analyses, only a few key risks are investigated, and by using this grid it should be easy to see which those risks are.

The probability impact analysis may be used for each option; however, it may also be used to identify those options that warrant further detailed quantitative analysis, saving time and effort in the analysis process.

Priority

Risks are given a priority based on their probability of occurrence, impact on occurrence and objective affected on its occurrence. This can be weighted system, which may ensure objectives not necessarily measurable on a monetary basis receive similar degrees of attention as to those that are.

During the appraisal phase, there are a large number of risks in the project, since few decisions have been made and there is a high level of flexibility. However, as the project progresses more decisions are made, which should reduce the amount of risk in the project, but also reduces the ability to make changes to the project, and increases the cost of making these changes.

The benefits from carrying out a risk analysis are reduced as the project progresses and the number of risks remaining is small. As the project nears completion there is likely to be very little change in the risk distribution and an exhaustive risk analysis can then cost more than the worth of the information that it produces. Therefore, the system proposed focuses upon the appraisal of projects where risk analysis has the greatest and most efficient means of influencing a project's outcome.

Risk evaluation

Evaluation of the base case model with risks both on a quantitative and qualitative basis determines specific strengths and weakness in relation to meeting the project's objectives. Stakeholders may have specific policies about the management of risk, and therefore clients may have to identify areas where guarantees or support may have to be provided when distributing risks throughout the various stakeholders.

Economic parameters

To assess the quantitative effects of risk upon the project, usually in accordance to the project cash flow, parameters such as cash lock-up, internal rate of return, net present value, payback period, debt service coverage ratios and return on equity may all be used to illustrate the implications of risks. Also, key stakeholders may stipulate hurdle rates with respect to the economic parameters. The evaluation of the base case may identify stress points within the cash flow, which acts as a dialogue for the identification of options when applying value and risk management.

Iso-risk curves

The assignment of probabilities and impacts can then be mapped against iso-risk curves, which dictates the level of attention particular qualitative risk demand. Those risks with a high priority are identified and addressed through the future development of project options.

In this case, four risks have been qualitatively identified with regards to their probability of occurrence and probable impact, Figure 4.6.

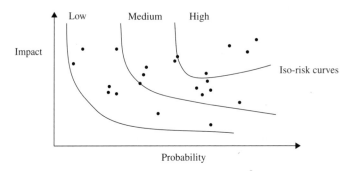

Figure 4.6 Plot of iso-risk curves.

4.7 Applying value and risk management

There are several value management techniques, such as re-engineering, business process re-engineering, and value planning that may be applied during the appraisal or development of solutions to a project. To save time and effort users must establish systems that are practically applicable, resulting in quantifiable outcomes. In this model *optioneering* – the identification and testing of several options to a problem, is a widely practiced approach when determining the viability of projects. In fact, it is a prerequisite for the submission of business cases for PFI/PPP projects in the United Kingdom.

Identification of options

From the evaluation of risk and the base case model established, a series of options that address the project objectives and risks both quantitative and qualitative may be identified. Options such as: do nothing (only applicable if there is a current asset already in operation), do minimum; refurbishment and/or rehabilitation and fast-track are common options to be included. However, options may also be named after specific project objectives, where the emphasis of the design, construction or operation of the project is geared specifically to address specific objectives. This becomes a mix between the qualitative vision or design philosophy behind the project and the quantitative costs and revenues generated thereafter.

Being aware of what produces best value and project success are critical to the identification of project options.

Analysis of options

Once the options are agreed, the original base case model and its risks are adjusted in terms of cost, time and quality. Both quantitative and

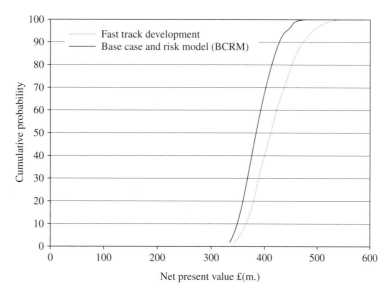

Figure 4.7 Cumulative probability distribution for the NPV of option A and B.

qualitative risk analysis are conducted to establish how each option performs. Variations in the economic parameters and probability impacts are derived. To be fair to each option, the techniques applied to each option should be maintained to protect the probity of any decisions derived.

Evaluation of the options: the VFM assessment

After analysis when all the options are presented for the project, comparisons between the various options may be completed. Specific options may present clear VFM based on differences in the economic parameters.

An example of the VFM assessment of a concession-based estuary crossing contract is discussed below.

A crossing of an estuary was quantitatively modelled using a network-based model. The two preferred options were a fast-track development (FTD) on site A, resulting in a higher land and planning costs, compared to the base case model, situated on site B.

As the fast-track development generates revenue earlier than the base case alternative, the net present value (NPV) is improved throughout the duration of the concession by £31 m. To complete the VFM assessment a 75% percentile envelope is formed, to illustrate the most likely outcome. From this assessment the FTD always offers greater VFM in terms of the total NPV achieved, from Figure 4.7:

(Million) Sterling £ (STE)	**15%**	**85%**
Base case risk model **(BCRM)**	354	424
Fast-track (FTD)	374	464
VFM of FTD	20	40

However, there may be instances where the quantitative VFM achieved is not so clear-cut. There may be instances where the two lines cross over and where under certain circumstances or the occurrence of specific risks one option out performs the other. Further investigation is therefore recommended, which may result in another option being proposed leading to a slight iteration of the value and risk management process.

4.8 Iteration of the process

The iteration of risk and value management assessments of a contract strategy should only occur during the appraisal of the project; however the project proposal will be iterated throughout the duration of the procurement process.

After the quantitative or qualitative analysis of the project, areas of weakness may be identified in relation to the risk and value management techniques' ability to portray or address the objectives of the project. Alternative techniques may have to be identified or created to allow for specific appraisal anomalies. The distribution of risk may also be untenable to the promoting/contracting agents, which may be identified during market testing or preliminary contract negotiation.

Adjustments made to the risk and value analysis techniques need to be recorded in the value plan, justifying why specific changes were necessary to support the decision making process. It is critical that the project team, do not appear to be moving the goal post, biasing the process.

Adjustments made to either the assessment of risks or the objectives held by the project should remain in the value plan to aid effective project and contract management.

Overall, the final phase should ensure that the option selected would meet the objectives developed in the first phase.

Clients should therefore re-evaluate:

❑ the validity of the original project objectives;
❑ identify the most promising option;

- identify the major risks associated with the option selected;
- assess the VFM of a contract strategy or project solution.

The second review is concerned with the means of achieving the project objectives as well as the objectives themselves. This should result in:

- a clear statement of the processes to be provided;
- a preferred outline design proposal;
- the basis for continued design development;
- a list of the risks associated with the design chosen;
- roles and authority of the project team;
- a proposal of how the project is to be implemented.

The client or sponsor, in association with stakeholders, will need to consider the risks associated with the project development. As the project moves from appraisal into contract negotiation, information will become available leading to the review of the option selected by the client. This is further encouraged by the submission of variant bids. Clients should attempt to actively engage the contractors to establish more efficient options, not necessarily considered during the initial project appraisal. The system adopted may be used as a means to test the validity of specific contract strategies as used in the formation of the public sector comparator for privately financed projects, or for the appraisal of project solutions. It is paramount that the risks associated with the final option are catered for in the contract strategy chosen.

4.9 Summary

Integrated risk and value management systems are used by project management departments to improve the likely success of a project by mitigating the risks and focusing on the value generative areas of the project. This chapter outlined an approach to an integrated management system for the management of risk and value during the inception and appraisal of a project.

Using an integrated approach to risk and value management can be challenging, but it is recognised as critical to the evaluation and maintenance of VFM. While there are several value management techniques available, this chapter focused on the application of optioneering in the appraisal of projects. Combining risk management techniques produces a more in-depth analysis of the project helping to deliver a successful project.

References

Merna, A. and Lamb, D.J. (2004) *Project finance: The Guide to Value and Risk Management in Public-Private Partnerships*, Euromoney PLC, London.

HM Treasury (2003) *The Green Book: Appraisal and Evaluation in Central Government*, HMSO, London.

HM Treasury (2004a) *Value for Money Assessment Guidance*, HMSO, Norwich.

HM Treasury (2004b) *Quantitative Assessment, User Guide*, HMSO, Norwich.

National Audit Office (2004) *Refinancing the Public Private Partnership for National Air Traffic Services*, The Stationary Office, London.

Chapter 5
Qualitative Methods and Soft Systems Methodology

This chapter illustrates the role of qualitative methods in risk management. Often the first stage in any assessment has to be a qualitative approach because there is insufficient information available to proceed with any quantitative methods. The value of a risk log is reviewed. Finally, the soft system methodology (SSM) is examined in detail and the way that it structures the investigation of the commercial situation.

5.1 Qualitative risk assessment

The first stage in any risk management process is also the first stage in the qualitative assessment of risk. It is frequently the most useful part of the risk management process and it lays the foundation for all the subsequent stages in that process, including the quantitative analyses that are frequently required to define budgets and timescales. Elsewhere in this book techniques are described which can be used to model the soft issues which influence projects, but the basic techniques for understanding risks and their potential influence are those of identification, assessment, ranking, sorting, classifying, allocating ownership and judging the probability and impact of potential risks. This is qualitative risk assessment. Frequently, no further analysis needs to be done. It is more likely that further analysis is firmly rooted in the qualitative process. Applying weighting factors to the qualitative assessment provides a quasi-quantitative form of analysis. Whatever the eventual outcome, the basis is the identification of potential risks. This process has been outlined in Chapter 4.

5.2 Review of project programmes and budgets

It is important that a project's programmes and budgets are realistic if it is to meet its objectives in terms of its quality and performance whilst remaining within its predetermined timescale and budget. Unless the budget and programme are achievable, it is unlikely that risk analysis will predict the out-turn cost and duration. This depends upon several factors including:

❑ the experience of the project management organisation;
❑ the amount of relevant data from closely similar projects which can form the basis of performance specifications, estimates and programmes;
❑ the extent of innovation; and
❑ the size and complexity of the project.

Budgets should be based on a realistic programme for the work taking into account resource provision, productivity, time-related costs and risks. Appropriate estimating techniques should be used for the type of project and the project stage at which the estimate is produced.

For example, broad-brush estimates prepared for the purposes of comparison between options during the appraisal stage of projects may be prepared using simple unit rate estimating methods or parametric methods (where suitable data exists). Estimates prepared to give definitive budgets for the selected option should be prepared using the operational technique taking into consideration the programme, resources and materials requirements of the project. Current costing can be applied to ensure that the estimate does not rely on updating of historical data.

The operational technique has the added advantage of facilitating greater understanding of the particular risks and uncertainties in the project and how they may impact on the project.

In the case of the budget, the review should ascertain:

❑ its adequacy for the scope of the works to be executed;
❑ any contingencies, allowances, provisional sums, etc. contained within it;
❑ the reason(s) for their inclusion;
❑ their adequacy; and
❑ the adequacy of elements related to overheads, supervision, consultants fees, licences etc. (if any) and any other realistic cost which may be identified.

The outline programme should be checked to ensure that:

❑ all the key activities have been identified;
❑ the durations are realistic; and
❑ the logic links and any other constraints are correct.

Such constraints may include, for example, the links to, or dependencies on:

- other projects;
- approvals for safety cases;
- approvals by statutory authorities (planning permission, etc.);
- approval of programmes on method statements; and
- approval of subcontracts and materials.

If the programme is in network form, the critical path(s), free and total float must be identified. All assumptions underlying the budget programme must be identified and logged.

Within each project, the following interfaces must be identified to ensure that they are included in the programme and managed effectively:

- between design groups;
- between design groups and specialists;
- between design and procurement;
- between design and construction;
- between procurement and construction; and
- with other projects.

Management will be facilitated by ensuring that each such interface is logged as a risk so that the following data are recorded and the following actions are undertaken:

- define data each party requires from others;
- define when they are required;
- agree assumptions if data is not available on time;
- log the assumptions;
- revise assumptions until final data is available;
- specify physical factors:
  - spatial positions;
  - loadings;
  - capacity, etc.;
- monitor progress; and
- achieve agreed dates.

Experience shows that frequently the project programme is in insufficient detail to identify all the areas of uncertainty, all constraints and all the interfaces. Hence, one of the key activities in the risk management process is to ensure that the programme is sufficiently detailed to fully understand all the activities that are required to execute the project.

5.3 The risk log

The results of the interviews and reviews of the programme and budget should form the basis for a risk log or risk register that will list all the identified risks. It will also contain assessments of their potential impact on the budget, programme and quality/performance aspects of the project.

To aid manipulation the risk log can be entered into a database system to facilitate recording, storing and sorting under various headings. These may include inter-alia:

- project phase;
- the owner (holder) of the risk;
- location;
- other use-defined categories, for example, cross references to the project programme and budget.

A database facilitates ranking of risks according to qualitative assessment (high, medium and low). It also permits quantitative estimates in terms of percent probability and cost impact (allowing quasi-quantitative analyses and ranking). Some database applications will allow quantitative analyses using Monte-Carlo simulation.

The output may be shown in the risk map (or matrix) format, also known as the probability/impact grid (PIG) to ease understanding of the results. In the case of quasi-quantitative analyses, cumulative frequency curves and histograms can be produced.

The risk log will also contain the information on actions to avoid, mitigate or transfer risks, the secondary risks arising and possible fallback plans. The risk log will be capable of being updated and will provide an audit trail. It is possible to use the risk log as a management tool to prompt risk owners to take action. Status reports can also be generated. An example of another risk log is shown in Table 5.1. It may be the case that for regular review by senior management, the full risk log or register is too detailed and cumbersome to be used as a management tool.

Given that senior management will be interested in the most significant risks, but are unlikely to be interested in details of modelling data, or the audit trail, it is possible to use an extract of the risk log that omits the detail but focuses instead on the action plans, the progress being made against them and their success in avoiding or mitigating risks. This should be reviewed regularly as part of regular management meetings. The team should be continually aware of any new risks, or those that are increasing in likelihood or severity.

Table 5.1   Main categories of risk.

❏   The project's constitution and organisational structure including the number of parties and the contractual, or other, relationships between them.

❏   The project management team including experience and availability of key personnel (in-house, consultants and contractors).

❏   Management authority and approvals required for work to proceed.

❏   Site-specific safety procedures: permits required, etc.

❏   Ground conditions, including special factors such as the extent of contaminated ground.

❏   Requirements of diversions, for example, to services.

❏   Risks arising from the contract/procurement strategy, including residual risks if the subcontractor does not perform.

❏   Risks arising from interfaces.

❏   Uncertainties and assumptions in the project scope/design.

❏   Temporary works for construction/dismantling.

❏   Potential for cost growth due to:
   ■   design development;
   ■   increased extent of identified risks such as contamination;
   ■   delays to approvals, permits, etc.;
   ■   delays due to contractor default;
   ■   unforeseen circumstances.

❏   Familiarity of potential contractors with the specific type of work and location.

❏   Extent of competition between potential contractors and suppliers.

❏   Delivery periods of materials and equipment.

❏   Extent, if any, of novel work.

❏   Constraints on the project programme, due to resource/staff shortages, possibly due to competing projects.

❏   Preventative measures to protect staff, labour and the surrounding areas.

❏   Special measures required for the handling and disposal of waste, spoil or contaminated material.

The reality of projects is that they do not follow a trend of steadily decreasing risk. Risks fluctuate in importance, and the reporting should reflect the changes.

Risks should be linked to the project's programme to understand the timeframe in which they can occur and the lead time to initiate preventative action. Whenever possible and cost effective to do so, risks should be avoided. In reality that must be mitigated as the project progresses. Many of these will be ongoing, that is to say they may span over several activities so that they cannot be closed for a considerable time.

5.4 Using a risk log to formulate risk management strategy

Following from the creation of a comprehensive risk log, an overview of the total likely risk exposure of the project can be formulated, based on the sensitivity of the budget and programme to identified risks and their potential impact in terms of budget overrun, delay and impact on the project's performance objectives.

The aim is to determine the most cost-effective strategy of risk avoidance, mitigation and/or transfer. The factors to be considered are:

- ❑ the potential impact(s) of each risk;
- ❑ the possibility of avoiding the risk through management action, provided that any secondary risks are not too great (secondary risks are those that arise as a consequence of taking the mitigating action);
- ❑ the possibility of taking actions to mitigate the risk, for example, by carrying out more thorough ground surveys to provide better information to the project team, its contractors and consultants. In this case, the risk to the probable cost of the ground works must be more than the cost of the survey otherwise the additional cost of the survey is not worthwhile;
- ❑ risks may be passed (i.e. transferred) to other parties, for example, a consultant, supplier, contractor or insurance.

In the case of risk transfer, two tests must be applied:

(1) Cost-effectiveness. It is usual for a premium to be charged by the party accepting the transferred risk. The issue is whether or not the premium to be paid is significantly less than the probable cost of accepting the risk in the first place. There is, however, a second consideration.

(2) The ability of the transferee to manage and accept liability for the risk should it occur. This is particularly important when significant risks to which the project is sensitive are passed to others, for example, contractors or suppliers. It may be that the risk premium charged by the contractor or supplier for accepting the risk is inadequate to

cover the cost of remedial action and the contractor or supplier is unwilling, therefore to carry it out. If this is the case, the project may suffer an adverse impact greater than the cost of retaining the risk would have been in the first place, bearing in mind that the premium paid to the transferee might be irrecoverable. (This is the residual risk of the contract procurement strategy.)

In other cases, the contractor/supplier may claim that the risk was excluded from his contractual responsibilities or was unforeseeable. When such claims are successful, the employer will effectively pay twice for the transfer of the risk:

❑   once through the premium charged by the contractor/supplier; and
❑   once through the successful claim.

It is particularly important therefore that the employer gives very careful consideration to risk transfer through contracts, the risk premium which contractors and suppliers are likely to charge and the types of contract available to achieve the optimum risk minimisation strategy. It must be borne in mind that in competitive bidding, contractors may not be able to fully price the risks that they are expected to carry. If the risk occurs, there is in fact no funding for its consequences or mitigation.

Whatever type of contract is chosen, it is essential that specialist contractors and suppliers who are best able to manage specific risks are used. The following factors must be considered:

❑   the extent of overlap of design, procurement and construction, if any, to achieve the desired completion date;
❑   transfer of risk and the premium(s) to be paid;
❑   transfer of control;
❑   transfer of responsibility; and
❑   the number of interfaces between contractors/suppliers which must be managed.

It must be noted that theoretical advantages may be difficult to achieve (e.g. price certainty through fixed price contracts where risk is high or the scope is not well defined).

The objective of any procurement strategy is to achieve the best VFM at the least risk. Fundamental to this is the understanding of realistic cost levels for tenders so that unrealistically low bids are not accepted.

In so far as risk assessment by the employer is concerned, a detailed understanding of the risks to be carried by contractors and suppliers, or to be shared with them, will enable:

❏ tender documents to be drafted to ensure that appropriate information is elicited from bidders;
❏ tender assessments to include a full appreciation of the risk being carried, how they will be managed by the bidders and what the implications are for the employer.

This is achieved by:

❏ ensuring that risks are identified and clearly specified in the tender documents;
❏ that the allocation of risks and responsibilities in the contract documents is clearly defined;
❏ the risk log can be used as a checklist during the tender assessment.

As noted above, even when risks have been passed to a contractor or supplier, there is the residual risk that they will not manage or will succeed in passing it back totally or in part through claims. Contingent sums should be allowed in budgets for these residual risks.

5.5 Qualitative methods

Commercial environments, particularly those where the management of risk is of prime importance, are frequently unstructured and only partly understood by those involved. Conflicting views are frequently held. The need to provide these environments with a structure should always be recognised at an early stage. Two principal means of achieving this structuring are available. First, the problem could be modelled using a prescriptive decision making tool. Second, a methodology that structured the investigation could be utilised that would permit the use of appropriate analytical tools. Thus, the choice is between a well-structured quantitative method and a project specific qualitative methodology.

Most of the qualitative techniques treat problems in environments where a single answer is assumed to exist, and the selection of an appropriate means to achieve an end that is defined at the start and thereafter taken as given. This approach is perfectly acceptable in the analysis of fixed facilities where the issues are purely technical. However, where human actions play a major role, or where uncertainty exists, these methods are inappropriate. There are arguments against the use of formal systematic models in favour of those methodologies that allow for solution of problems that cannot be fully structured in advance. The former of these approaches creates artificially scientific environments in which the problem is depoliticised, people are treated as passive objects and uncertainty is ignored. The assumptions made are often based on abstract objectives from which concrete actions are proposed for implementation.

Qualitative methodologies concern themselves with how management decisions are actually made, rather than the traditional operational research approach of obtaining the *right* answer. Methodologies that can screen out unfeasible alternatives, study the entire range of solutions, and explore the effect of likely constraints, will develop contrasting possibilities as to what is required. Placing decisions in the context of alternative future environments permits the opening up of discussions about threats and opportunities. Simplicity and clarity are sought, and uncertainty treated as a fact of life. People are treated as active subjects. Outside influences such as technical, commercial and political considerations are identified and considered in direct relation to internal issues. Appropriate strategies are developed to deal with complex interactions within the project. The methodologies utilise a *bottom-up* approach and facilitate participation by those directly involved in the problem. They are non-optimising, and accept that there will be future uncertainties and options should be kept open wherever possible.

A number of methodologies to tackle complex problems have been developed over the past two decades. These include the analysis of inter-connected decision areas (AIDA), conflict analysis, robustness analysis, strategic options development and analysis (SODA), the strategic choice method and the soft systems methodology (SSM). The purpose of these methodologies is to structure inquiry into situations that are characterised by uncertainty, conflicting objectives and significant human involvement.

These methodologies fall into two broad classifications. First, are those that concentrate on the efficiency of the solutions proposed (the first three methodologies listed above falling into this group). Second, are those that concentrate on the effectiveness of the solutions (as represented by the last three methodologies listed above). While these methodologies concentrate on one of the two criteria, they do appreciate the other criteria, rather than excluding it as most multiobjective methods do. Of these, SSM concerns itself with systematically desirable and culturally feasible changes, rather than simply making better decisions. For these reasons, SSM was chosen as the methodology to employ in the investigation of a water company's activities as described in the following section.

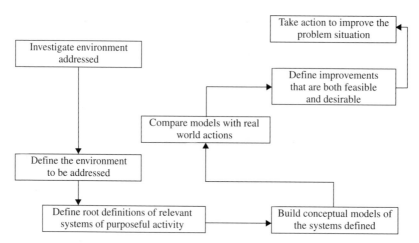

Figure 5.1 The learning cycle of SSM.

5.6 Soft systems methodology

SSM was developed by Peter Checkland at Lancaster University in the late 1970s and early 1980s. Its purpose was to overcome the inability of traditional decision theory and to adequately solve not all but the most structured problems. The importance of these structured problems is usually far less than those involving uncertainty and social considerations. A particular strength of the technique is that it can begin with a simple desire to *make things better*. No definition further than this is required.

SSM is typically employed in a cycle of seven stages, as indicated in Figure 5.1. The first two stages involve finding out about the situation considered problematic. The first two stages investigate the environment and culture in which the problem exists, the specific problems considered, the reasons why the situation is considered problematic and the improvements that are sought in the third stage of SSM. A view of the problem is selected that provides an insight into how improvements can be achieved. This is undertaken through the use of root definitions and neutral definitions of the activities or task to be undertaken that provide insight into the problem.

The fourth stage involves the building of conceptual models that are logical expansions of the root definitions generated in the previous stage. The models developed are those of systems that can adapt and survive to changes through their processes of communication and control. The fifth stage of SSM requires the comparison of the models developed with reality. This provides a means of instigating debate into how benefits in the systems can be attained. This process directs attention on the assumptions made, highlights alternatives and provides an opportunity for re-thinking many aspects of real-world activity.

The purpose of the sixth stage of SSM is to define changes that will bring about mediation benefits. Such changes have to meet the criteria of systemic desirability and cultural feasibility. Systemic desirability will include such factors as creating mechanisms to determine effectiveness and ensuring that logical dependencies are reflected in real-world sequential actions. Cultural feasibility will provide allowance for the illogic of human actions and the political environment in which decisions are taken.

The final stage of SSM is taking action and implementing the changes proposed. Undertaking the proposed changes alters the perceptions of the initial problem. If required, further cycles of SSM can be employed to seek additional improvements. This process would have been made considerably more straightforward through the structuring of the problem undertaken in the first application of SSM.

The use of the CATWOE mnemonic is a means of increasing the understanding of the problems considered and the ease with which conceptual models can be developed. The CATWOE mnemonic represents the *C*ustomers, *A*ctors, *T*ransformation process, *W*eltansschauung (world view), *O*wners and *E*nvironmental constraints on the problem.

The core of CATWOE is the linkage of transformation processes and the world view that makes them meaningful. Any activity will always have a number of transformations by which it can be expressed, dependent upon the perception of its purpose. The other elements in CATWOE add the ideas that purposeful activity must be undertaken, that it could be stopped, that there will be victims and beneficiaries and that this system will take some environmental constraints as given.

SSM employs a number of assumptions. These include that SSM is a process for management, and is a means of achieving purposeful action to obtain a change in an existing situation. In any given situation there will be conflicting issues and remains from the parties involved in that situation. SSM makes use of systems ideas and treats situations holistically.

Systems are considered to be composed of natural activity systems that can be linked together in a logical structure. As a number of possible descriptions for a given situation will always exist, it is imported to be explicit about the view taken of the problem to be studied, the *Weltanschauung*. The next assumption made is that SSM learns by comparing models of purposeful action with perceptions of the real world. This provides a feedback mechanism to determine the efficacy, efficiency and effectiveness of the proposed actions. The final assumption made is that the process must be participative so that all those parties involved make an input, even if they are not aware of the methodology or models employed.

To date, SSM has been used in a wide variety of public and private organisations. The applications include performance evaluations, educational studies and the appraisal of various commercial situations.

5.7 Case study: SSM in the use of the placement of construction projects

Having provided an overview of SSM, an account of the hypothetical commercial situation considered, the procurement of an ongoing series of projects by a UK utility is now presented to provide a summary of the last two stages of SSM. There are a number of ways in which UK utilities can procure construction projects, and several organisational structures that can potentially be employed. These include retaining all design and management in-house, employing consultants for these roles, and design and build contracts. The hypothetical utility considered, procured between five and ten of the specialised projects considered annually, and employed the first of these options, the method considered in this case study. The investigation of the problem, the first stage of SSM, was undertaken investigating the structure, processes and climate of the organisation.

Projects considered were procured in support of one of the utility's core business activities, the majority of projects originated from one of the utility's core business areas or from the company's planning arm. An established database of completed projects existed. This database contained all the tenders received over the past three years and records of expenditure during the construction of the projects, including variations, risk events, post-contract reviews and audits.

A list of pre-qualified specialist contractors was invited to tender for projects. Cost based target contracts are employed. The contingencies employed were tailored to the individual project rather than applied uniformly to all projects, and took account of the ability of the contractor employed. The final process undertaken by the utility was the aggressive marketing of their expertise in this field.

The utility was committed to promotion of specialist construction techniques as a means of reducing the cost of construction. This commitment was expressed through publicising successes, educating internal and external clients and maintaining as high a profile of the utility as possible.

Close relationships existed between the utilities and the contractors employed. The procurement mechanisms employed, particularly the use of cost based contracts, allowed the company to make significant inputs to the construction management of the projects without incurring cost penalties.

Competition is considered to exist within the utilities to procure the projects. The use of cost based contracts can erode the competitiveness of the tenders received from contractors. Thus, whilst the construction division procured projects from a logical and historical perspective, other business areas occasionally retained projects if they thought that they could obtain significantly lower tenders, and hence total project costs. In many cases, however, the tenders obtained by other business areas bore little relation to the final costs.

The third stage of SSM involves identifying those areas considered problematical. The list of problems below is probably not exhaustive.

Technical and environmental

(1)   Uncertain ability of construction technique to adequately deal with anticipated and unforeseen ground conditions.
(2)   Difficulties in transporting soil away from construction sites.
(3)   Problems in detecting existing services.
(4)   Difficulties in achieving economic means of disposing of slurry in an environmentally acceptable manner.
(5)   Interfacing with existing utilities and obtaining possessions.

Commercial and operational

(6)   Inability of the construction technique to compete on cost terms with traditional forms of construction in the majority of cases.
(7)   High cost of purchasing and maintaining specialist plant.
(8)   Inability of client organisations to guarantee long-term workload of projects.
(9)   Contractors attempting to buy work in order to obtain future work from the client.
(10)   Inability of clients to motivate contractors in the long term in accordance with their short-term aims.
(11)   Continued scepticism of the construction techniques by large sectors of UK industry.
(12)   Inability of clients to motivate contractors to perform in accordance with their short-term aims the minimisation of costs.
(13)   Uncertainty over means of paying contractors to minimise construction costs.
(14)   Uncertainty over the effectiveness of incentive mechanisms, either positive or negative.

(15)   Perceived high risk-nature of construction technique.
(16)   Even minor risks cause major cost and time overruns.
(17)   Difficulty in knowing cause and hence allocation, of construction risks.
(18)   Adversarial relationships created when traditional conditions of contract were employed through their inability to allocate risk equitably.
(19)   Risk averse behaviour of contractors through their perceived use of excessive contingencies.
(20)   Inability to differentiate contractors based on their ability to manage construction risks.
(21)   Difficulty in predicting the productivity of specialist plant.
(22)   Inability to differentiate contractors based on the efficiency with which they operate their plant.

Of the problems identified, the most fundamental was the inability of the construction technique to compete on cost terms with traditional construction techniques. Where clients are unable to provide a guarantee of long-term workloads of projects, contractors face uncertainty over the number of projects over which they could write off the cost of plant. If the technique is not adopted as the construction technique for a given project, none of the other problems can occur.

The next most important problems considered are the inability of clients to differentiate between contractors based on their ability and their inability to motivate contractors in accordance with their long- and short-term aims.

The abilities of contractors varied significantly in terms of their ability to manage construction risk. Where the differences between competing tenders were lower than single figure percentages, it is difficult for the company to justify the employment of more expensive contractors. These differences in contractors' abilities are influenced by a number of factors, including their size, the duration they have been in existence and the financial requirements of their stakeholders. The financial and operational problems identified above are either indirectly or directly related to the procurement process.

Root definitions

The next stage of SSM is the formulation of root definitions. This requires the naming of systems considered relevant to exploring the problems identified in the exploratory phase of SSM. The following root definition and CATWOE mnemonic were produced; root definitions give neutral definitions of the activities or tasks to be undertaken. For the purposes of the following root definitions, a root definition and CATWOE mnemonic of the activities based on the finding-out phases of SSM, are presented below.

Root definition:

> An internal organisation, seeking to employ improved systems to procure special projects for internal and external clients to increase the cost competitiveness of the technique compared with traditional construction methods.

C (Customers):	Utilities, external clients, contractors.
A (Actors):	Utility.
T (Transformation processes):	Construction needs of clients met through increased procurement of specialist construction projects through the greater cost competitiveness.
W (World view):	The increased cost competitiveness of the special construction techniques that can achieve significant financial and non-financial benefits.
O (Owners):	Utility.
E (Environmental constraints):	Water companies capital programme, UK construction market.

The CATWOE mnemonic indicates that the transformation sought is one in which the cost competitiveness of the specialist construction techniques is increased. This increased competitiveness is achieved through the use of procurement mechanisms tailored to the requirements of the projects. A fundamental assumption is that the procurement of the projects is based on an ongoing workload of projects work rather than on an individual basis.

Figure 5.1 indicates the schematic structure of the procurement activity system. This system contains the elements through which microtunnelling projects are currently procured. The system appreciates the direct construction and promotional roles and activities of the utility.

In order to propose systems through which improvements could be obtained, the problems were structured according to the general areas that they address. For this reason, problem numbers 7–11 were considered to be procurement problems occurring at the strategic level. The remainder of the problems were considered at the tactical level in that they are applicable to individual projects.

Problems 12–14 considered the abilities of the payment mechanism employed to reimburse contractors. Problems 15–20 considered the management and allocation of construction risks, and problems 21 and 22 concerned the contractor's operation of their plant. Four activity systems were proposed which address these classifications of problems. The root definitions and associated CATWOE mnemonics of these systems are presented below.

Root definition 1: procurement of microtunnelling projects

Root definition:

> A system, appreciative of the construction environment, to procure projects accordance with the client's long-term aims.

C:	Utility, contractors.
A:	Utility contractors.
T:	Projects procured. The procurement of projects improved.
W:	The use of procurement systems tailored to the requirements of the project and the contracting parties will improve the effectiveness and efficiency of the procurement process.
O:	Utility.
E:	Legislation, UK economy, utility capital programme.

Root definition 2: payment of contractors

Root definition:

> A system to pay and motivate contractors to perform in the short term in accordance with the client's aims.

C:	Utility, contractor.
A:	Resident engineer, site agent utility client contractor, construction staff.
T:	Contractors paid. Contractors paid and motivated to perform in direct accordance with the client's long-term aims.

W:	The use of financial motivators can have a major effect on construction performance.
O:	Utility.
E:	Local construction markets.

Root definition 3: management of construction risk

Root definition:	A system in which construction risks are allocated between the client and the contractor to minimise the construction cost of an ongoing workload of projects.

C:	Contractors, utility.
A:	Utility.
T:	Workload of projects constructed. Projects constructed at lower cost.
W:	The effective management of construction risks can have a major effect on construction cost.
O:	Utility.
E:	Current/local construction market climate, client's capital programme.

Root definition 4: operation of construction plant

Root definition:	A system to ensure contractors' effective operation of plant and provide a mechanism for differentiating between contractors on the basis of their technical ability.

C:	Contractors, utility.
A:	Contractors.
T:	Contractor's operation of their plant. Contractor's operation of their plant monitored and improved.
W:	The ability of the contractor to operate their plant efficiently can have a major effect on the productivity achieved.
O:	Utility.
E:	Individual project conditions.

The first root definition, concerning procurement at the strategic level, seeks to overcome the problems associated with the use of specialised construction techniques at the organisational, rather than at the construction level. Thus, this root definition considers long-term relationships and contractual arrangements between the contracting parties at the organisational level. The second root definition, describing a system to reimburse contractors, assumes that the payment mechanism employed can motivate contractors to achieve the client's aims.

The third root definition, describing a system to manage construction risks, takes the world view that the effective allocation of construction risks can have a significant effect upon construction costs. The system seeks to allocate construction risks so as to reduce the cost of constructing an ongoing workload of projects. The contractors' operation of their plant is the system described by the fourth root definition.

The management of construction risks exists at the lowest level of the procurement process of the activity systems considered, and is the system addressed further here. If the improvements sought in the procurement of projects are to be effectively utilised, they must have a consistent and coherent hierarchy. If this were not the case, elements of the procurement process could impose conflicting demands and aims on the contracting parties. This is consistent with the *bottom-up* approach employed by SSM.

Figure 5.2 provides a view of the first risk management system component. The left hand branch of this component appreciates the effects of risk on the construction projects (box a), and the construction environment in which the projects are procured (box b). The term construction environment is used to represent the commercial and organisational climate in which contractors are employed to construct the projects. The linking of these elements provides an understanding of the specific effects of risk on the construction of an ongoing workload of projects.

The left hand branch of this component introduces the analytical aspects of risk management. Box c first identifies the locations of primary risk data. Construction risks are categorised according to their source, and their durations incorporated into a database of risk events. The risk analysis alternatives are reviewed in box d in terms of the findings sought and the primary risk data available. When the two branches of this component are combined, a mechanism for calculating the financial effect of utilising alternative risk allocation strategies is obtained (box e). Finally, the development and verification of methods for calculating the effect of employing alternative risk allocation strategies, the fifth stage of SSM, is undertaken, indicated by box f.

This structured approach permits the assessment of a range of risk allocations strategies, using primary data, to simulate the impacts upon the constructing parties.

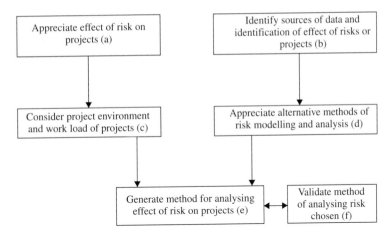

Figure 5.2 Identification, assessment and analysis of construction risks, the first component of the risk management system.

Proposed risk allocation strategy

The risk allocation strategy proposed represents the sixth stage of SSM, defining improvements in the situation. The aim of this strategy is to minimise the construction cost of the portfolio's projects. The client accepts responsibility for all risks caused by unforeseen natural and artificial conditions. The responsibility for all the other classifications of construction risk is shared between the contractor and the client on a 50/50 basis.

The responsibility for delays due to natural and artificial conditions is a frequent source of dispute between the client and the contractor. The reasons for this include uncertainties over the encountered conditions, even after risk events have occurred. These risks have low probabilities of occurrence, but high probabilities of effect, meaning that contractors face a gamble when setting contingencies against them and have at best limited abilities to manage them. This inevitably increases contractors' tenders.

The third reason is the need to provide contractors with motivation to seek to eliminate the occurrence of risks. The level at which this motivation becomes most effective is clearly difficult to identify. However, given the manner in which even minor risk event can have major financial effects, the contractors' reduced exposure to risk will still provide them with an incentive to perform in accordance with the client's aims.

The risks allocated to the contractor are those that they have the ability to manage; they are not allocated based on expediency, as occurs in price based contracts. It is cheaper in the long run for the client to pay for what actually happens during construction rather than what the contractor thinks might happen. The arbitrary transfer of responsibility for risks to the contractor costs money, whilst their effective allocation and management can save money. Removing the responsibility for these risks altogether diminishes the contractors' motivation to operate effectively, meaning that risk events will be more likely to occur. The proposed strategy reduces contractors' responsibility for risk provided they seek to minimise their likelihood of risks occurring. Although the contractor will still estimate the cost of these construction risks higher than the client, the construction costs will be reduced.

It has been proposed on a general basis elsewhere that the client accepts a larger responsibility for construction risks. Where major unexpected events do occur, the client will inevitably face increases in construction cost unless a contract placing all risks on the contractor is employed. However, if the risk was logically the client's in the first place, the cost to the client will be no greater that it would otherwise have been. The approach proposed requires the adoption of a long-term perspective. The workloads of projects are not constructed concurrently. Thus, the time period considered is not just the construction duration of an individual project, but several years. The client must accept a flexible attitude towards risk allocation. This approach would not be suitable for a client with a more introspective culture or fewer projects.

The use of cost based contracts is a prerequisite for the successful use of this strategy as a precise knowledge of the contractor's actual costs is required. The IChemE Green Book contract, already widely used, would be an appropriate vehicle for this strategy. The Institution of Civil Engineers (ICE) engineering and construction contract, or New Engineering Contract (NEC3), would also be suitable, particularly as they have provisions for sharing and transfer of risk between the contracting parties once identified thresholds are met.

The cost of some projects will inevitably exceed the client's estimates, particularly when major unforeseen events occur. However, as the client takes a long-term view these cost overruns should be more than offset by the cost savings achieved on successfully completed projects. The contingencies utilised by the contractor will have the same purpose and effect as those currently used in IChemE cost based contracts. The benefits are shown at portfolio level, where contingencies to cover the impact of the major risks at project level can be held.

In addition to minimising construction costs, a further aim of this strategy is to promote good management and engineering practice by the client as well as the contractor. These aims are compatible. The effect is to have the construction aims of the client and contractor identical. A *win–win* environment is sought in which the minimisation of construction risks and hence costs, is in the best interests of both parties. The contractor's risk becomes their ability to obtain further projects from the client. This approach raises the contractor's risk from the project level to the long-term success of their business, the contractor becoming as good as their last project.

Before such a radical departure from existing practice is embarked upon, a process of validation must be undertaken. The behaviour of the model is tested within the feasible domain to ensure that it is compatible with real life and that it contains no discontinuities or step changes.

There are a number of validation alternatives available, including a comparison with historical and future projected data, and comparing the simulated output with the anticipated outcome of a project in response to a predefined set of parameters. To validate this research a number of prototype contracts would have to be procured and analysed.

5.8 Summary

The initial role and value of qualitative analysis has been reviewed and the use of a risk log investigated. Finally, this chapter has shown that SSM provides a suitable means for structuring the examination of a utility's procurement of an ongoing workload of construction projects. The actual procurement of these projects is assumed to be undertaken by an internal specialist organisation. Root definitions of systems have been proposed indicating areas in which improvements in the use of special construction techniques can be obtained.

The allocation of construction risk, contractor payment mechanisms and strategic procurement mechanisms were identified as the aspects to be investigated. It is not suggested that these are the only areas through which improvements could be obtained. However, these are the areas where it is thought that meaningful improvements can be obtained given the resources and constraints present.

The proposed risk allocation strategy promotes the employment of able contractors, and it has been shown that only these contractors are able to obtain their required returns if the client seeks to limit their contingencies and profits in exchange for guarantees of work.

Chapter 6
Quantitative Methods for Risk Analysis

Project appraisal or feasibility study is an important stage in the evolution of a project. It is necessary to consider alternatives, identify and assess risks, at a time when data is uncertain or unavailable. This chapter outlines the quantitative approach and describes in detail several risk assessment techniques.

6.1 Sanction

When the project is sanctioned, the investing organisation is committing itself to a major expenditure and is assuming the associated risks. This is the key decision in the life cycle of the project. In order to make a well-researched decision the client will require

Clear objectives The client's objectives in pursuing this investment must be clearly stated and agreed by senior management early in the appraisal phase for all that follows is directed at achievement of these objectives in the most effective manner. The primary objectives of quality, time and cost may well conflict and it is particularly important that the project team know whether minimum time for completion or minimum cost is the priority. These are rarely compatible and this requirement will greatly influence both appraisal and implementation of the project.

Market intelligence This relates to the commercial environment in which the project will be developed and later operated. It is necessary to study and predict trends in the market and the economy, anticipate technological developments and the actions of competitors.

Realistic estimates/predictions It is easy to be over-optimistic when promoting a new project. Estimates and predictions made during appraisal will extend over the whole life cycle of implementation and operation of the project. Consequently single figure estimates are likely to be misleading and due allowance for uncertainty and exclusions should be included.

Assessment of risk   A thorough study of the uncertainties associated with the investment will help to establish confidence in the estimate and to allocate appropriate contingencies. More importantly at this early stage of project development, it will highlight areas where more information is needed and frequently generate imaginative responses to potential problems, thereby reducing risk.

Project execution plan   This should give guidance on the most effective way to implement the project and to achieve the project objectives, taking account of all constraints and risks. Ideally, this plan will define the likely contract strategy and include a programme showing the timing of key decisions and award of contracts.

It is widely held that the success of the venture is greatly dependant on the effort expended during the appraisal phase preceding sanction. There is however, conflict between the desire to gain more information and thereby reduce uncertainty, the need to minimise the period of investment and the knowledge that expenditure on appraisal will have to be written off if the project is not sanctioned.

Expenditure on appraisal of major engineering projects rarely exceeds 10% of the capital cost of the project. The outcome of the appraisal as defined in the concept and the brief accepted at sanction will however freeze 80% of the cost. The opportunity to reduce cost during the subsequent implementation phase is relatively small, as shown in Figure 6.1.

6.2 Project appraisal and selection

Project appraisal is a process of investigation, review and evaluation undertaken as the project or alternative concepts of the project are defined. This study is designed to assist the client to reach informed and rational choices concerning the nature and scale of investment in the project and to provide the brief for subsequent implementation. The core of the process is an economic evaluation; based on a cash flow analysis of all costs and benefits that can be valued in monetary terms.

Appraisal is likely to be a cyclic process repeated as new ideas are developed, additional information received and uncertainty reduced, until the client is able to make the critical decision to sanction implementation of the project and commit the investment in anticipation of the predicted return.

It is important to realise that, if the results of the appraisal are unfavourable, this is the time to defer further work or abandon the project. The consequences of inadequate or unrealistic appraisal can be expensive, as in the case of the Montreal Olympics stadium, or disastrous.

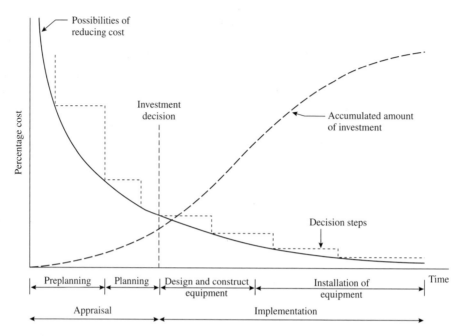

Figure 6.1 Graph of percentage cost against time showing how the important decisions for any project are made at the start of that project.

Ideally, all alternative concepts and ways of achieving the project objectives should be considered. The resulting proposal prepared for sanction must define the major parameters of the project – the location, the technology to be used, the size of the facility, the sources of finance and raw materials together with forecasts of the market and the predictions of the cost/benefit of the investment. There is usually an alternative way to utilise resources, especially money, and this is capable of being quantified, however roughly.

Investment decisions may be constrained by non-monetary factors such as:

❑ organisational policy, strategy and objectives;
❑ availability of resources such as manpower, management or technology.

Programme

It will be necessary to decide when is the best time to start the project based on previous considerations. Normally this means as soon as possible, because no profit can be made until the project is completed. Indeed, it may be that market conditions or other commitments impose a programme deadline, that is, a customer will not buy your product unless it can be supplied by mid-1998, when a processing factory will be ready. In inflationary times, it is particularly important to complete a project as soon as possible because of the adverse relationship between time and money. The cost of a project will double in 7.25 years at a rate of inflation of 10%.

It will therefore be necessary to determine the duration of the appraisal design and construction phases so that:

- the operation date can be determined;
- project costs can be determined; and
- the client's liabilities can be assessed and checked for viability. It may well be that the client's cash availability defines the speed at which the project can proceed.

The importance of time should be recognised throughout the appraisal. Many costs are time-related and would be extended by any delay. The programme must therefore be realistic and its significance taken fully into account when determining the project objectives.

Risk and uncertainty

The greatest degree of uncertainty about the future is encountered early in the life of a new project. Decisions taken during the appraisal stage have a very large impact on final cost, duration and benefits. The extent and effects of change are frequently underestimated during this phase although these are often considerable, particularly in developing countries and remote locations. The overriding conclusion drawn from recent research is that all parties involved in construction projects would benefit greatly from reductions in uncertainty prior to financial commitment.

At the appraisal stage, the engineering and project management input will normally concentrate on providing:

- realistic estimate of capital and running costs;
- realistic time scales and programmes for project implementation;
- appropriate specifications for performance standards.

At appraisal, the level of project definition is likely to be low and therefore risk response should be characterised by a broad-brush approach. It is recommended that effort should be concentrated on:

- seeking solutions that avoid/reduce risk, however care is needed to ensure that the consequences of avoiding risk, the secondary risks are not worse than the original risk;

❑ considering whether the extent or nature of the major risks is such that the normal transfer routes may be unavailable or particularly expensive;

❑ outlining any special treatment, which may need to be considered for risk transfer, for example, for insurance or unconventional contractual arrangement;

❑ setting realistic contingencies and estimating tolerances consistent with the objective of preparing the best estimate of anticipated total project cost; and

❑ identifying comparative differences in the riskiness of alternative project schemes.

Construction project managers will usually have less responsibility for identifying the revenues and benefits from the project: – this is usually the function of marketing or development planning departments. The involvement of project managers in the planning team is recommended, as the appraisal is essentially a multidisciplinary brainstorming exercise through which the client seeks to evaluate all alternative ways of achieving these objectives.

For many projects, this assessment is complex as not all the benefits or disbenefits may be quantifiable in monetary terms. For others it may be necessary to consider the development in the context of several different scenarios (or views of the future). In all cases, the predictions are concerned with the future needs of the customer or community. They must span the overall period of development and operations of the project that is likely to range from a minimum of eight or ten years for a plant manufacturing consumer products to 30 years for a power station and much longer for public works projects. Phasing of the development should always be considered.

Even at this early stage of project definition, maintenance policy and requirements should be stated, as these will affect both design and cost. Special emphasis should be given to future maintenance during the appraisal of projects in developing countries. The cost of dismantling or decommissioning may also be significant but is frequently conveniently ignored.

6.3 Project evaluation

The process of economic evaluation and the extent of uncertainty associated with project development are illustrated by the appraisal of the hypothetical new industrial plant in Chapter 7. The use of a range of financial criteria for quantification and ranking of the alternatives is strongly recommended. These will normally include discounting techniques but care must be taken when interpreting the results for projects of long duration.

Cost–benefit analysis

In most construction projects, factors other than money must be taken into account. If a dam is built it might drown a historical monument, reduce the likelihood of loss of life due to flooding, increase the growth of new industry because of the reduced dam flooding risk, and so on. Cost–benefit analysis provides a logical framework for evaluating alternative factors that may be highly conjectural in nature. If the analysis is confined to purely financial considerations, it fails to recognise the overall social objective, to produce the greatest possible benefit for a given cost.

At its heart lies the recognition that a factor should not be ignored because it is difficult or even impossible to quantify it in monetary terms. Methods are available to express, for instance, the value of recreational facilities, and although it may not be possible to put a figure on the value of human life, it is surely not something we can afford to ignore.

The essential cost–benefit analysis is to take into account all the factors, which influence either the benefits or the cost of a project. Imagination must be used to assign monetary values to what at first sight might appear to be intangibles. It should be mentioned that monetary values are highly subjective and must be evaluated with care. Even factors to which no monetary value can be assigned must be taken into consideration. The analysis should be applied to projects of roughly similar size and patterns of cash flow. Those with the higher cost–benefit ratios will be preferred. The maximum net benefit ratio is marginally greater than the next most favoured project. The scope of the secondary benefits to be taken into account frequently depends on the viewpoint of the analyst.

It is obvious that, in comparing alternatives, each project must be designed within itself at the minimum cost that will allow the fulfilment of objectives including the appropriate quality, level of performance and provision of safety.

Perhaps more important, the viewpoint from which each project is assessed plays a critical part in properly assessing both the benefits and cost that should be attributed to a project. For instance, if a private electricity board wishes to develop a hydroelectric power station, it will derive no benefit from the coincidental provision of additional public recreational facilities, which cannot therefore enter into its cost–benefit analysis. A public sector owner could quite properly include the recreational benefits in its cost–benefit analysis. Again, as far as the private developer is concerned, the cost of labour is equal to the market rate of remuneration, no matter what the unemployment level. For the public developer however, in times of high unemployment, the economic cost of labour may be nil, since the use of labour in this project does not preclude the use of other labour for other purposes.

6.4 Engineering risks

An essential aspect of project appraisal is the reduction of risk to a level that is acceptable to the investor. This process starts with a realistic assessment of the uncertainties associated with the data and predictions generated during appraisal. Many of the uncertainties will involve a possible range of outcome that could be better or worse than predicted.

The implications of several of the risks likely to be encountered in engineering projects are illustrated in Figures 6.2(a–e). It is relevant to note:

❑ that the single line investment curve shown in Figure 6.2 represents the *most likely* outcome of the investment. An idea of the spectrum of uncertainty arising from the estimates and predictions is shown in Figure 6.2(e);
❑ the maximum risk exposure occurs at the point of maximum investment – when the project is completed and either does not function or is no longer needed.

Figure 6.2(b) uses the dotted lines to illustrate the impact of both a greater and lower level of income generation on the project cash flow.

Figures 6.2(c) and (d) shows the significance and sensitivity of the cash flow to a delay in completion date and a delay in sanction data respectively.

Risks specific to a project are interactive, sometimes cumulative: they all affect cost and benefit.

Environmental risks frequently result in compromise following comparison of cost with benefit. They are likely to have a significant influence on the conceptual design and the response should therefore be agreed prior to sanction. Residual uncertainty may be incorporated in the analyses, usually as a contingency sum that may have to be expanded.

Risk to health and safety is normally considered as a hazard during design and embraces issues such as reliability and efficiency in addition to safety. In the case of facilities that process hazardous substances

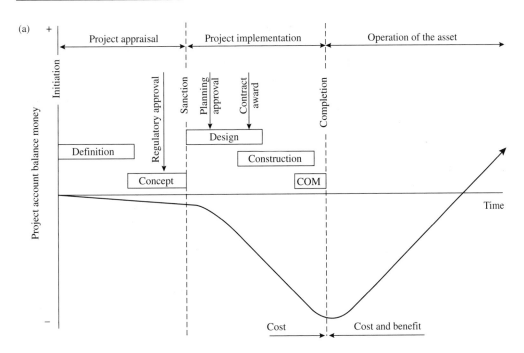

Figure 6.2 (a) Project cash flow – typical cash flow.

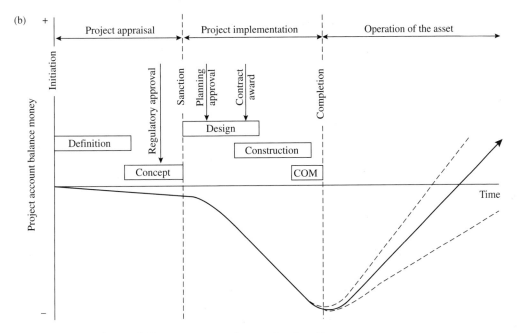

Figure 6.2 (b) Project cash flow – spectrum of operational performance.

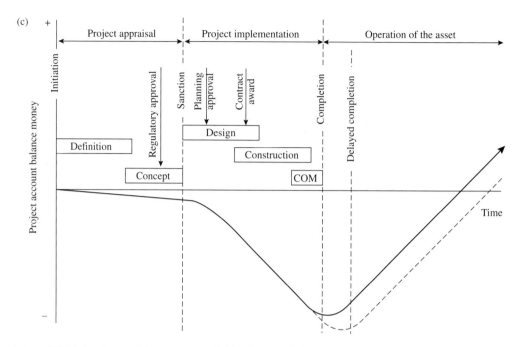

Figure 6.2 (c) Project cash flow – effect of delay in completions.

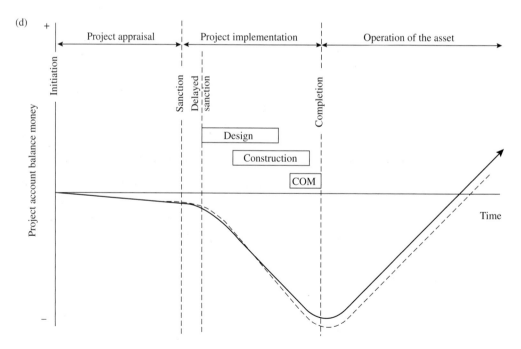

Figure 6.2 (d) Project cash flow – effect of delay in sanction.

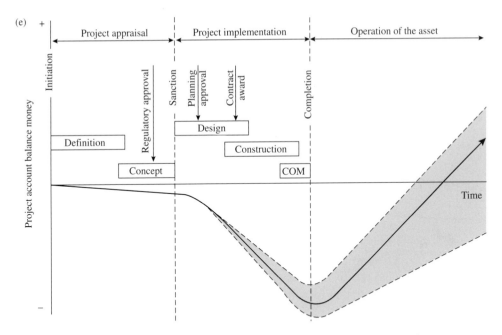

Figure 6.2 (e) Project cash flow – spectrum of change in implementation and operation.

a full-scale safety audit will be necessary or mandatory. This will have implications for both programme and cost.

Innovation The consequential risk of inadequate performance may be reduced by thorough testing but appropriate time and cost provision must be included.

Risk to activity relates mainly to the implementation phase of the project. These risks arise mainly from uncertainty and is the responsibility of the project manager who should be allocated appropriate contingencies. The extent and nature of the contingencies depends on the magnitude and complexity of the risks and on the degree of flexibility required.

All uncertainties, particularly those that cause delay, will affect investment in the project. Many risks are associated with specific time constraints imposed on the project. The preparation of an outline programme is an essential early requirement of any approach to risk identification.

6.5 Risk management

The logical process of risk management has been defined earlier as:

❑ identification of risks/uncertainties;
❑ analysis of the implications (individual and collective);

❑ response to minimise risk;
❑ allocate appropriate contingencies.

If uncertainty is managed realistically, the process will:

❑ improve project planning by prompting what if questions;
❑ generate imaginative responses;
❑ give greater confidence in estimates;
❑ encourage provision of appropriate contingencies and consideration of how they should be managed.

Risk management should impose a discipline on those contributing to the project, both internally and on customers and contractors. By predicting the consequences of a delayed decision, failure to meet a deadline or a changed requirement, appropriate incentives/penalties can be devised. The use of range estimates will generate a flexible plan in which the allocation of resources and the use of contingencies are regulated.

Risk reduction

❑ Obtaining additional information.
❑ Performing additional tests/simulations.
❑ Allocating additional resources.
❑ Improving communication and managing organisational interfaces.

Market risk may frequently be reduced by staging the development of the project. All the above will incur additional cost in the early stages of project development.

Contingencies

The setting and management of contingencies is an essential part of project management. The three types of contingencies are: time (float), money (allowance in budget) and performance/quality (tolerances).

Their relative magnitude will be related to the project objectives. The responsibilities/authority to use contingencies should be allocated to a named person. It is essential to know what has been used and what remains at any point in time.

The role of people

All the above risks may be aggravated by the inadequate performance of individuals and organisations contributing to the project.

Control is exercised by and through people. As the project manager will need to delegate, he/she must have confidence in the members of the project or contract team and, ideally, should be involved in their selection.

Staff should be involved in risk management in order to utilise their ideas and to generate motivation and commitment. The roles, constraints and procedures must be clear, concise and understood by everyone with responsibility.

6.6 Probabilistic analysis

For real projects, ranges of values are preferable to single figure estimates because of the level of optimism that is inherent in a single figure estimate. Probabilistic risk analysis techniques are used to provide information such as estimates of the likelihood of achieving certain project targets and the likely range of outcomes of the project, in terms of its duration and economic parameters. There are a number of different probabilistic risk analysis techniques and each technique requires the specification of key project variables and their corresponding distributions. A probability distribution is used to describe the ways in which a value may be selected as representative of the estimated range of outcomes of a variable.

Probabilistic analysis techniques require a large number of calculations to be carried out, and this usually requires the speed and processing power of a computer. There are a number of computer programs available for use in the risk management process and most of these programs utilise one of the probabilistic analysis techniques described in this section.

Monte-Carlo technique

The Monte-Carlo technique was so called because of its imitation of the randomness of a roulette wheel. This technique was developed a number of years ago and has become one of the most popular probabilistic risk analysis techniques. Computer programs often make use of this technique in conjunction with model simulations. It has been well documented and the full mathematical calculations and equations required can be found in a number of texts.

The Monte-Carlo technique is a process for developing data using a random number generator. It should be used for problems involving random variables with known or assumed probability distributions. This technique requires the selection of different values from a probability distribution, the values corresponding to their probability of occurrence as defined by the probability distribution. In the analysis phase, the identified risks are quantified. The quantitative risk is usually included in the risk model by estimating a pessimistic, a normal and optimistic value, known as a triangular distribution, although others can be used (see below). It is very important that the risk analyst manage to transfer the information gathered in the identification phase into risk assessments reflecting the real risk affecting the parameters. Often the spread is far too conservative, that is, the risk is underestimated. It is also very important to discuss the assumptions behind the estimates to avoid the risk assessments that are anchored to the estimates.

It is very important to use a practical and approximate approach when quantifying risk and selecting probability distributions. Do not turn your project into a complete mathematical equation. Keep it simple!

Examples of typical risk distributions are given in Table 6.1. The risk quantifications as shown in the table were allowed for an industrial project and are based on subjective judgements (experience and knowledge) and *gut feeling* of members of the project team and is used as input to a risk model.

A simple explanation as to the way in which a Monte-Carlo technique operates is now given. A value is chosen from the probability distribution of each of the variables and run through the project model. Each pass through the project is called an iteration. It is normal to carry out one thousand, or more, iterations for an average project to ensure that the results are free from most statistical biases.

The Monte-Carlo technique requires a sequence of random numbers that have no predictable pattern and satisfy various statistical tests of randomness. Output histograms produced from a Monte-Carlo analysis can be tested in two ways to see if they are *robust*. The histograms should be sensitive to a change in the number of iterations used and insensitive to a change in the random number initiator.

Table 6.1   Examples of simple risk distributions.

Activity	Optimistic	Planned	Pessimistic	Distribution
Construction	16 m $	20 m $	30 m $	Lognormal
Equipment	8 m $	10 m $	12 m $	Triangular
Market size	35 m people	50 m people	65 m people	Triangular
Market share	6%	12%	36%	Triangular
Unit price	1 $	2 $	4 $	Triangular
Operational cost	1.6 m$	2 m $	2.4 m $	Triangular

Note: m = million.

However, the use of random numbers implies that all of the variables are independent of each other, and in many projects, this is not true. It is often the case that variables are interdependent, and a common example of this is when there is a delay in the design stage this often leads to a delay in the construction of a project. Sometimes interdependent project variables are specified as correlated in order to overcome this assumption.

The Monte-Carlo technique does not require the analyst to have a great amount of knowledge about computer modelling or statistical risk analysis techniques in order to use it effectively. However, it is not advisable to use it on projects where there are significant number of interdependent variables, unless the model specifies that, the variables are interdependent. Typically, the number of key variables would be less than 10% of the elements or activities of a project model.

When using a probabilistic analysis technique it is necessary to choose a probability distribution that is representative of the range and the way in which the values of a variable might vary. There are a number of different shapes of probability distribution that can be chosen to represent the probability of occurrence of a range of values that comprise a variable. The shape of the probability distribution chosen is often based on historical data, which would show the distribution of the outcomes for this variable on past projects. The use of historical data is an objective means of deciding the shape of the probability distribution for a variable. However, historical data from which to choose the shape of a probability distribution is not always available and in this event, it is left to the project modeller and those associated with the project to decide on the probability distribution shape. In addition, it can perpetuate old problems!

There are four commonly presented probability distributions, the normal, beta, rectangular and triangular distributions although the triangular (or triple) estimates are the most often used. This is because on most projects, definitive information regarding the likely distributions of key variables is not available but what information exists is likely to reflect a *most likely* outcome and an indication of the relative optimistic and pessimistic range of outcomes. In practice, not all distributions are symmetrical; it is often the case that probability distributions are skewed because it is more likely that there will be a delay in completing an activity than the activity being completed early.

It is important to choose a probability distribution that is appropriate for the variable being considered. If there is no historical data available, then careful consideration should be given to the variable and what is likely to be the result of that variable, before a decision is made about the shape of the distribution. Utilisation of the experience of members of the organisation, from previous similar projects, might assist in the decision as to which probability distribution is most appropriate for a particular variable.

6.7 Response to risks

Since all projects are unique and risks are dynamic through the life of the project, it is necessary to formulate responses to the risks that are appropriate. The information gained from the identification and analysis of the risks gives an understanding of their likely impact on the project if they are realised. This, in turn, enables an appropriate response to be chosen.

Typically, there are three main types of responses to risks, and these are – to avoid or reduce the risks, to transfer the risks and to retain the risks.

Risk avoidance or reduction is an obvious first stop. Once the risks, particularly the sources of risks, have been identified and analysed, it may be possible to formulate methods of avoiding certain risks while making only minor changes to the project. In extreme cases, projects may be abandoned due to an inability to avoid or reduce some of the risks. However, there will only be a few occasions when this response can be used, because the project can only be changed to a certain extent before it becomes either infeasible or unviable or becomes a different project!

By changing certain features of the project, it may be possible to reduce the amount of risk in the project, rather than trying to avoid the risks totally. It might be possible, for example, to change the method of construction to reduce the amount of risk involved in the construction phase of the project, while making little impact on the duration and cost of the project.

Risk transfer involves transferring risks from one party to another, without changing the total amount of risk in the project. Risk transfer can occur between the parties involved in the project or one party and an insurer. The decision to transfer or allocate risk to another party is implemented through an insurance policy or the conditions of contract. It is usually up to the client to initiate the transfer of risk, although there are several factors that need to be considered before any risk is transferred. First, consideration should be given as to whether or not the party that the risk is being transferred to, can do anything to manage or control the risk, and whether they could accept the consequences should the risk be realised. It is generally agreed that risks should be accepted by the party that is best able to manage or control them, or the party that is best able to accept the consequences should they be realised. There is little point in transferring a risk to a party that cannot manage the risk or cannot accept the consequences should it be realised. The second consideration is whether or not the risk premium that would have to be paid for the transfer of a risk is greater than the cost of the consequences should the risk be realised. Again, there is usually little point in paying more to transfer a risk than it would cost to accept the consequences should the risk be realised. Equally, there could be the problem that a low tenderer has not priced risk at all.

In some situations, the only option available is to retain a risk. The party that is holding a risk might be the only one that can manage the risk or accept the consequences should the risk be realised. The risks retained may be controllable or uncontrollable. If the risks are controllable, then control may be exerted to reduce the likelihood of occurrence or the impact of the risks. It is normal for the client to be left with some risks, and these are termed the residual risk. The client ultimately carries the risk in a project.

The timing of an action taken to mitigate the effects of a risk may dictate the action that is chosen. The first possible action is on advice that reduces the chance of the risk being realised. Then there are after-the-fact actions using readily available resources. This refers to an action taken once the risk has been realised, using the resources that are available at the time the action is taken, a purely after-the-fact action requiring essential prior actions. This action requires the use of contingency measures that are planned prior to the start of the project.

Despite the use of formalised procedures and techniques in the risk management process, the element of human judgement must not be overlooked. One of the most significant uses of judgement is in deciding which actions should be taken to manage the risks. There are no rules that can be applied in this situation, it is up to each party in the project to decide whether they are prepared to accept a risk or whether they wish to find some means of avoiding or transferring the risks but this depends upon the thoroughness and clarity of the risk identification and analysis. Judgement in this process will be influenced, to a certain extent, by the decision-makers perceptions of the risks and their attitude toward risk taking.

6.8 Successful risk management

Risk management is undertaken by both client and contractor organisations, but for different reasons. Clients will usually be concerned with the best use of their capital resources, the likely cost of procuring the facility and their return from their capital investment. Contractors will be concerned with the decision as to whether to tender for a given project in terms of the returns obtainable, the desired competitiveness of their tender and the most profitable means of constructing or increasingly designing and building the project.

The duration that clients typically take to propose, appraise and sanction projects can range from a number of months to many years. Contractors are often given a matter of weeks to tender for projects. It is apparent that the risk management exercise undertaken by the client should be more comprehensive than that undertaken by the contractor. In addition, clients often take the view that the sort of risk that might cause a contractor to default is exactly the type of risk that they do not wish to carry.

It is important to ensure that the requirements and nature of the project concerned determines the mechanisms employed, rather than selection based on expediency. There would be little point in employing risk management methods that require, for example, precise risk data if none was available but this is often the case.

The prerequisites for successful risk management would seem to be: full specification of the project and all identified associated construction risks, a clear perception of the construction risks being borne by each party, sufficient capability, competence and experience within the contracting parties to manage the identified construction risks and the motivation to manage risks, requiring a clear link between a party's ability to manage and actual management of risks and their receipt of reward.

The effectiveness of risk management is improved if all parties to a contract have the same appreciation of the identified risks. The contractor and the client should have similar views of the likelihood and potential effects of all risks. This can be achieved if pre-contract discussions between the client and the contractor ensure that a clear, mutual understanding of the relevant risks. A lack of understanding may lead to the contractor under-pricing their tender and pursuing claims for additional sums as the project proceeds. Alternatively, contractors may deliberately price low and expect to recover money through claims.

6.9 Principles of contingency fund estimation

Contractors employ contingencies in an attempt to guarantee their return when construction risks for which they are responsible occur. These contingencies represent the risk premium that the client pays for allocating construction risks to the contractor. The risk allocation strategy employed can have a major effect on the contractor's tender for a project, particularly when the project is perceived as high-risk by the contractor.

Current risk modelling techniques allow good correlation between the risks identified and quantitative risk analysis (QRA). Current risk management practise uses the risk register as a key management tool to identify risks and track management responses.

A danger inherent in combining the risk register and the QRA is that there is a belief that because discrete risks are identified, logged and modelled and then if a risk has not occurred, the value assigned to it can be subtracted from a budget figure that corresponds to a specified confidence level.

The errors in taking this view are outlined below:

❏ The input data for risk analyses are subjective and approximate. If precise data exists then by definition the issue in question cannot be a risk. It follows that the results of risk analyses are approximations. They are not accurate in the accounting sense.

❏ The model uses a sampling technique so the contribution of each risk to the total at any given confidence level is not known. If a risk has been avoided the reduction in the estimated cost should be calculated from running the model again with the risk deleted. The effect of deleting a risk must not be calculated by subtracting the value of the risk (itself a subjective evaluation) from a total budget at a given confidence level.

❏ Whilst it is correct to link certain risks to activities in the programme, it is an oversimplification to treat the programme as timelined that identifies dates by which risks should occur. First, the programme is at risk so any date should in fact be a range of dates. Second, although the programmed date that the risk should have occurred may have passed, the risk could have occurred without being reported. For example, a contractor may only notify the occurrence of a risk when he claims compensation for it some time after the event itself. Third, many risks are not discrete events that can be linked to a single point in the programme, so it is difficult to say they should have occurred by a given date.

The fundamental point is that the results of QRAs cannot be treated as entries in a set of accounts. Expressing in another way, the risk register does not constitute a budget with line items.

When the use of risk management as a separate discipline in project and construction management was being developed in the late 1970s and early 1980s it was recognised that many of the key risks that could be identified were the soft issues that are very difficult to quantify with precision. It was obvious that to ignore these risks because they were imprecise and difficult to quantify is absurd. It was equally obvious that the traditional approach to making allowances for these risks, plus anything else that could be thought of or happened to come along later, by adding a single figure contingency of say 10%, is inadequate. This approach does not define risks as discrete items. This means that they are frequently overlooked by management and not quantified.

To improve the situation and allow more realistic plans and estimates to be prepared, risks are identified and computer simulations are used to model complex scenarios. Put very crudely, whenever we prepare an estimate or plan, we usually consider more than one value: an optimistic outcome and a more pessimistic outcome, then we choose a value somewhere in between. When the single figure is selected, two values out of three are discarded. That is to say, more than two-thirds of our knowledge is discarded. In fact it is worse than this because the optimistic value may not be the best possible and the pessimistic is not the worst that could occur. In other words, neither the opportunities nor the major risks are included.

The main purpose is to demonstrate that there is a range of possible outcomes for a project rather than a single value and to show how risk and uncertainty influences that range. It was never the intention, nor is it possible, to treat the results of risk analyses as accurate forecasts of future outcomes. They are better approximations than single figure estimates with a nominal contingency sum added. They also lead to better understanding and management of risk.

Annexe: Alternative methods of risk analysis

Portfolio theory

Portfolio theory was originally developed within the context of a risk-averse individual investor who was concerned with how to combine shareholdings in several different companies in order to build up an investment portfolio that would maximise his expected returns for a given level of risk.

At any given time, most organisations are involved in a number of projects, each project containing different levels of risk. When the various levels of risk for each of the projects are combined the organisation can find itself exposed to very high levels of risk. An organisation that does not consider the amount of risk to which it is exposed may easily overstretch itself by taking on too many *risky* projects.

Appraisal techniques tend to consider individual risks or the risks related to a single project and this can have serious implications for an organisation. Some practitioners have applied this theory to the portfolio of projects built up by an organisation. This enables the organisation to perceive the amount of risk to which it is already exposed, and to decide whether it would be able to undertake another project, in terms of the additional risks that the organisation would be exposed to. Portfolio theory is a theory that assists the organisation in choosing what can be termed the *efficient set* of projects.

Delphi method

The Delphi method is an established technique for obtaining consensus estimates from several experts, and this technique can be applied to the assessment of risks. The general procedure for this technique is that an estimate of the variable(s), or risks, is obtained from each of the experts. This estimate can relate to the probability of occurrence or the likely impact of a variable. The experts are then informed of all the estimates and asked to give a revised estimate. This process continues until a consensus estimate is produced. This method can be viewed as a qualitative or quantitative technique, since the experts may or may not be asked to provide a quantitative estimate of a variable.

The Delphi method can be adapted for use in the assessment of risk in projects. This procedure starts with the formation of a team of experts that represent all aspects of the project. This is an interdisciplinary team of experts formed for the purpose of assessing the risks in a project. The experts meet and formulate an exact definition of the risk that is being considered. They then discuss the risk, paying particular attention to its causes and the interdependencies it has within the project. These experts then give their opinions as to the probability of occurrence of the risk and the impact of the risk on the project should it occur. The experts can also give a cost assessment of the risk based on the probability of occurrence and possible impact. This procedure is based on the consensus of opinions of the experts involved. The procedure proposed differs from the classical Delphi method in that the opinions of the experts are not gained from a survey but instead the experts are drawn together in meetings presided over by a moderator.

This risk analysis technique is expensive, in terms of the resources used, the cost of the resources and the time taken. The technique relies heavily on the opinions of people deemed to be experts. If the group of experts chosen does not represent a sufficiently interdisciplinary team then the results produced may be biased and of little use. In order to produce results of any value the team must hear the opinions of experts from all fields related to the project. However, it would be difficult to find a time when all the experts could meet to discuss the risks in a project. If the classical Delphi method is used then it is not necessary to get all the experts together at once, thus allowing a wider range of experts to be used, but the time taken to get a consensus view would be significantly increased.

The Delphi method is a very subjective technique and the results gained from this should be viewed with caution. This technique would be best used on projects where there is little information available or where the organisation concerned has little previous experience of carrying out similar projects. Due to the expense incurred from using this technique those constrained by a tight budget should not use this technique.

Influence diagrams

Although this is a relatively new technique, it appears to be based on the much older network planning technique. The influence diagramming technique involves mapping out the project, identifying the sources of risk and possible responses to these risks. This information is then represented diagrammatically.

Although influence diagrams are essentially a qualitative method of analysing the risks in a project, costs and times can be included in the diagram if desired. To use this technique it is necessary to have some understanding of risk sources and their importance. However, the main advantage of influence diagrams is that the relationships between the risk sources and activities in the project can be easily seen. By being able to see these relationships it makes it much easier to identify effective responses to the risks, and in some cases, it is possible to identify one response to a number of risks.

This technique is very useful and relatively cheap, in terms of the time that it takes to perform the analysis and the resources that are required. The influence diagramming technique requires consideration of the entire project and then displays this information in a simple and understandable way. This technique assists in identifying risk responses that apply to several risks. However, it is a subjective technique and if used on projects that can be divided into a number of small sections the diagram becomes unclear. This technique is best applied to projects that are divided into a few major activities, where alternative strategies are being considered and where quantitative assessment of the risks is not required.

Decision trees

Decision trees, also known as decision networks, are diagrams that depict a sequence of decisions and chance events, as they are understood by the decision-maker. The decision tree is made up of two types of nodes, decision nodes and chance event nodes. A decision node represents a decision that has to be made and a chance event node represents an event that has a chance of occurring, possibly a risk. A decision tree starts at a decision point node on the left hand side and the information is conveyed going across the page from left to right. At the time represented by a specific node all prior decisions, or decisions to the left of the node, have been made and uncertainties related to prior chance event nodes have been removed. Each decision node should have at least one branch, or arrow, coming from it and these branches represent the decision alternatives.

The branches of a decision tree indicate the alternative courses of action that can be taken, and in this form, the decision tree can be considered a qualitative risk analysis technique. However, if probabilities are assigned to the branches of the decision tree indicating the likelihood of each course of action occurring or being taken, then the decision tree is used as a quantitative risk analysis technique. A decision tree does not necessarily have to have probabilities of occurrence assigned to the branches when it is being used as a quantitative technique; there are several other measures that could be used. Examples of other measures are the cost of taking a particular route or the gain expected from taking a route, and it is up to the decision-maker to determine which measure is most appropriate for the project.

The procedure for constructing a decision tree begins with the identification of decision points in the project and the possible alternative courses of action available at each of the decision points. Once this has been completed, it is necessary to identify the chance-event points, or uncertainties in the project and establish the possible alternative outcomes of each chance-event. When used as a quantitative technique, the quantitative information, such as the cost of the possible alternative courses of action, must be estimated. Finally, the decision tree should be evaluated to obtain the expected values for following each alternative course of action.

The main advantage of using a decision tree is that, whether it used as a qualitative or quantitative technique, it requires the entire project to be set out in a logical sequence. This ensures that the decision-maker has considered all the options available in the project at an early stage. An advantage of using decision trees is that they clarify and communicate the sequence of events to be considered in making a choice. Since decision trees are a diagrammatical representation of project information, they are easily understood. They give everyone involved a common understanding of the way in which the decision-maker perceived the project. This technique does not identify the best alternative or course of action to be taken; it merely sets out all the possible alternatives. If it is used quantitatively then it can give some measure of the likelihood of alternatives or courses of action occurring, or of the possible gain from taking a particular course of action.

Decision trees are a very useful technique for getting information across to those involved in the project. They are cheap and easy to produce, since they only require the use of one person who has a good understanding of the project, the chance events and the alternatives available. However, if the project has a large number of decision nodes or chance event nodes then the decision tree can become complicated. If it is used for a quantitative analysis on that type of project then the calculations involved become time consuming and tedious, and to some extent, subjective. The measure that is used in a quantitative analysis gives the outcomes derived from taking different routes, however, this measure is subjective and not always very meaningful. In projects that contain a number of chance event nodes the measure produced for each route can show the probability of that route actually being taken, but the probability of each route being taken is likely to be small due to the large number of opportunities to take different routes. This technique is best used to evaluate different approaches to a project. It is a good technique for communicating information.

Latin Hyper-Cube sampling

Latin Hyper-Cube sampling is a technique that statisticians recommend for use when there are a large number of parameters to be varied. For problems containing several variable parameters there will be a very large number of possible choices, and, in these cases, the sample size is usually smaller than the number of possible choices. In a case such as this, a Latin Hyper-Cube can be constructed and the sample is then chosen from this in a deterministic, statistical way. The values, once chosen, are then input into the model in same way as for the Monte-Carlo technique, producing a range of possible outcomes for the model. This is only a simple explanation of the technique, but information that is more detailed can be found in other texts.

The difference between this technique and the Monte-Carlo technique lies in the choice of the sample, therefore, many of the advantages and limitations are the same. Latin Hyper-Cube sampling is a relatively new technique, particularly in the field of project and risk modelling, giving a limited scope for its current use.

Chapter 7

The Contribution of Information Technology to Risk Modelling and Simulation

The real price of computers continues to fall and a wider group of people are computer literable and have computer access. This has lead to the increased use of computers, especially desktop and personal computers, in the process of risk management. Computers are fast and efficient tools for evaluating data but it is important that the users should not lose sight of the assumptions on which the software packages are based. The idea that if the computer has produced something then it must be right is a belief held by many people and is certainly not true. The output from a computer model is determined by the information input, which means that accurate data is essential, but the output is only part of the basis on which risk management decisions can be based.

The purpose of this chapter is to examine the role of computers in the risk modelling and simulation process, and to consider the types of risk management software (RMS) available. The distinction between the terms modelling and simulation are explained, and the advantages and limitations of the use of computers are assessed. The data requirements for realistic modelling and the information output from the various software packages are also considered.

This chapter also provides guidance for the new or inexperienced risk modeller. It covers the reasons for modelling projects, explains the pitfalls and the advantages and assists the modeller in developing their first model, with the use of a case study. Thus providing the new/inexperienced modeller with a sound basis from which to start, assisting them in constructing a mathematical model; and aiding them in understanding the outcomes of the models.

The reason for modelling projects is to gain an understanding of what might happen throughout the life of the project. It provides the opportunity to test some possible situations and gain an understanding of the way in which the project will react to different situations. Project models are an approximation of reality, and the first thing to understand is that no model is perfect, but as skills in modelling increase the modeller creates a more realistic approximation of reality.

7.1 Purpose of RMS

RMS forms one component of the suite of project management software packages that are commercially available. Undertaking risk management activities provides organisations with an understanding of the potential finance and resource commitments necessary to achieve the completion of a particular project or projects. RMS packages are most commonly based on the network technique and the critical path method, whose theoretical basis goes back to the 1950s. The other two principal components are planning packages and estimating packages. These packages have achieved wider use than RMS for a number of reasons. These include the ease of use and analytical simplicity of most planning and estimating packages and their wider applicability. Until recently, RMS was only considered necessary when projects perceived as being high risk were being appraised. This remains the case in many instances, although the increasing evaluation of projects over their entire lifetime and a focus on political and financial risks over technical risks has promoted the use of RMS.

Project management involves the sanction, construction and on occasion, operation of construction projects. This may place demands on the project manager not encountered in line management where essentially repetitive activities are undertaken. The demands of project management mean that the requirements for accurate and comprehensive information on which to base decisions are vital. In extreme cases where projects are subject to incalculable uncertainty, environments in which change can take place rapidly are required, rather than plan to deal with specific identified events. RMS aids the decision maker in these processes, primarily at the appraisal stage, but also throughout the project life cycle.

As project management programs become more and more a part of the everyday life of the project manager, it becomes important for the project manager to possess knowledge and skills in the areas of computing, simulation and simulation methods, statistics and mathematical modelling concepts in addition to the traditional qualifications needed for successful project management.

7.2 When to use RMS

The question to be addressed is when one should employ RMS packages. There are a number of situations when the effort and resource expended in the training and use of RMS is justified. Where a project is subject to risk and uncertainty from a number of dependent or independent sources, the effect of which is difficult to identify, RMS is often the only means by which the combined effects of these risks can be determined. These effects can be quantified using sensitivity and probabilistic analyses. Only in the most simple of cases can these analyses be undertaken without the help of RMS, and where very simple cases exist it is unlikely that these analyses would be required in the first place. In addition, correlations between risks can be incorporated when there is sufficient evidence to justify their use.

Where the project considered is constructed and operated over a number of years and in particular is being used to generate a number of revenue streams, the time-value of money is very important. In these cases, the modelling of the project using RMS can simplify and structure the appraisal of the project. *What-if* analyses are frequently required to determine the effect of modifying individual project and risk parameters. Where the project has to be remodelled a number of times, the effort required can be significant if the analysis has to be repeated manually. The use of RMS permits these analyses to be undertaken in an extremely simple manner. When a number of dissimilar projects are competing for funding, the use of RMS can provide a consistent financial basis from which to evaluate the projects.

The list of reasons for using RMS provided below is far from exhaustive, but does illustrate those considered most common.

❑ Modelling projects reduces the dependence upon raw judgement, and requires the analyst to employ a structured and rigorous approach to problems.
❑ Risk management software can carry out complex, iterative calculations faster than can be done either by pencil and paper exercises or by using standard spreadsheet packages.
❑ *What-if* analyses can be undertaken with a minimum amount of additional effort.
❑ Data can be modelled in a flexible manner, allowing the computer model to replicate the actual project.
❑ Outputs can be provided quantitatively and graphically, increasing the ease of understanding the results obtained.

However, the use of RMS does have disadvantages, which should not be underestimated. These are considered to include the following:

❑ The model will always be used on assumptions, some of which may be flawed.

- ❏ Undue belief can be placed on the results, rather than challenging them to a significant degree.
- ❏ The model produced is only a mathematical representation of real life and therefore may not accurately reflect the reaction of the actual project to real life complications.
- ❏ The selection of the most appropriate RMS package for a particular application is not always clear, and the evaluation of alternatives may take a considerable amount of time.
- ❏ Whilst the cost of many RMS packages has reduced considerably over the past decade, the time required by analysts to learn fully the various capabilities of RMS can be considerable, particularly for the more complex packages. Clearly this will result in the analysts' employer incurring costs during this learning process.
- ❏ Too much complexity that adds no value.

7.3 Requirements of the analyst

An appreciation of the mathematical processes employed by RMS is vital if their outputs are to be interpreted correctly. A lack of understanding of these processes can result in the selection of inappropriate RMS for a particular application.

If a sophisticated RMS is employed to model a project, it is important that the analyst have an understanding of, for instance, Monte-Carlo simulations and Markovian logic. The user need not have a detailed knowledge of the mathematics involved, but an awareness of the effect of altering parameters and variables is vital if the outputs of the RMS are going to be interpreted in the correct light.

In addition to knowledge of the RMS employed, the analyst must also have an appreciation of the project being modelled. A lack of familiarity with the technical and financial aspects of a project may result in the analyst making oversimplistic assumptions and failing to sufficiently capture the characteristics of the project during the modelling process. However, it is unlikely that a particular organisation would train large numbers of its staff in the use of RMS. Thus the analyst must also have a broad appreciation of their organisations business activities. The balancing of this knowledge of the RMS and the organisation's business is vital.

7.4 Modelling and simulation

The terms modelling and simulation are frequently used when discussing the use of computers in risk management process; however, the distinction between these two terms is not so frequently understood. In broad terms, modelling is the process of describing the project in a mathematical way, and simulation is the process of imposing real life complications on the model and measuring the effects.

A model can be regarded as an approximation to reality. A model can never be a perfect representation of reality as the actual reaction of the project to certain factors is unknown. However, it is important that the model is as realistic as possible to ensure that it will react in a similar way to the actual project when the real life complications are added, balanced with the complexity of the model which will increase its' development time and level of interpretation of outputs required.

Simulation involves the testing and experimentation with models, rather than with the real system. The models employed are representations of an object, system or idea in some form other than that of the entity itself. The computer is a powerful aid to this activity, and this is a reason for the recent increase in simulation-based methods.

Within the sphere of project management, one is principally concerned with the time and cost effects of investment decisions and management actions and mathematics is the best language in which to express these parameters. The third principal project parameter, quality, can usually only be expressed in qualitative terms. Thus quality is not included in risk management software.

A mathematical model will include constants, variables, parameters, constraints and mathematical operators. Their purpose is either optimisation or description. Optimisation will seek to identify the course of action that either maximises return or minimises expenditure; the purpose of the descriptive approach is to provide insight into the project considered.

The *Influence diagram* method is an intuitive method for decision analysis and risk management. It was formalised as a methodology in the early eighties. An influence diagram is simply a diagram, which consists of nodes reflecting variables and decisions, and influence is reflected by arrows.

The influence diagram method is a tool for reflecting both complex and simple problems. The major advantage is that the method allows for creative thinking and increased insight through graphical problem formulation. A model can cost not only time and cost estimates influencing the project goal, but also risk factors such as organisation, management and decision making, laws and regulations, political risks and financial risk.

The use of the influence diagram method makes it easier to formulate the problem and capture the opinion of experts. The diagram also provides a convenient way of expressing the nature of a problem to others and thus aids the general understanding of the factors that influence the outcome. Relatively complex relationships can be simply formulated, thus allowing the behaviour of these relationships to be studied in a way that would be impossible by pure reasoning.

Combining influence diagrams and the Monte-Carlo technique results in a very powerful tool for risk analysis. Several software packages exist on the market with this combination. The influence diagram method is a very flexible way of building the risk model, and it allows you to add all risks to your model, not only those that affect your time and cost estimates. The modelling processes also forces the project to consider and measure the effect of external risks that is usually much more important than the uncertainties, which are a natural part of the estimates anyway.

7.5 Modelling using RMS

If sophisticated RMS is being employed, it is important that there is a level of consistency between the methods and models being employed. For this reason, the simple three-step approach to risk management is often inadequate. An approach that has been shown to provide a better understanding of the modelling processes involved is given below:

- ❏ *Problem definition* Recognising and defining a problem to study that is amenable to analysis.
- ❏ *System conceptualisation* A concept model is a concise, systematically organised statement of the process, including the specifications of the inputs, outputs, processes involved and the parameters, variables and constraints employed.
- ❏ *Model representation* The stage where the model is presented to the risk management software in an appropriate form.
- ❏ *Model behaviour* Computer simulation is used to determine how all of the variables within the model behave over time.
- ❏ *Model evaluation* Tests are performed on the model to assess its quality and validity.
- ❏ *Policy analysis and model use* Finally the model is employed to test alternative policies that might be implemented.

The advantages that this approach brings include the formalising of the appraisal of the project. This forces the analyst to structure the appraisal in a sequential and logical manner that requires all assumptions made are being explicitly declared.

7.6 Data management

It is important that the data employed, particularly the data on which the risk parameters and variables are based, is treated in a coherent manner, consistent with the development of the model. Risk data can be obtained from three sources – past projects, ongoing projects and the opinions of individuals who have experience of the project under appraisal. The first two sources, considered as primary data, are likely to be more reliable than the third source. However this data is frequently unavailable, and where it is the effort needed to obtain it may be considered excessive. Thus the third of these sources of risk data, subjective opinions, is the basis from which risk variables and parameters are usually developed.

Consider the following example: an analyst estimates the likely individual effects of a number of risks as contributing to increase the cost of construction of a project by up to 10%. These figures are incorporated into a project model created using a RMS package. A risk analysis is performed, and the cost of the project at the 75% level of probability, 30% internal rate of return (IRR) is calculated as £1 500 000, and this figure is employed for future appraisals. In reality this figure is meaningless as there is no way that a single point estimate based on a number of subjective judgements could accurately reflect the actualities of constructing the project. However, if the results were supplied in terms of an S-curve with the gradient of the 15–85% quartile indicated, as shown in Figure 7.1, it would show a range of IRR from 27–35% that would provide a means of going forward.

This example also provides arguments against the initial or automatic use of correlations, variances and some of the more sophisticated statistical operations that can be employed during risk analyses. Including these elements for the sake of producing a sophisticated model merely dilutes the accuracy of the outputs obtained. These elements should only be employed where a database of risk data has been developed which can be sampled from, to generate statistically significant risk parameters and variables.

7.7 Analytical mechanisms

All of the RMS currently available employs some form of sensitivity and probabilistic risk analyses based on Monte-Carlo simulations, with the network packages employing Markovian logic to simulate the interdependence of project activities to the identified risks.

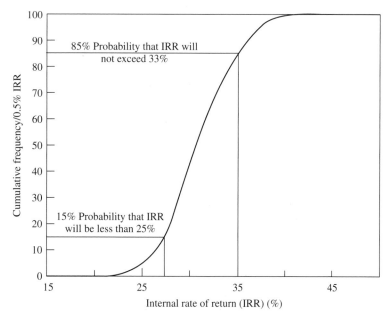

Figure 7.1 Cumulative frequency diagram – internal rate of return.

Sensitivity analysis provides answers to *what if* questions by isolating key variables and evaluating the effects of incremental changes in the values assigned to the key variables. Sensitivity analyses can pinpoint the most critical areas of a project in terms of their reaction to the occurrence of risks. The principal limitations of this technique is its inability to provide an indication of the outturn cost of completing a project, and when changing a variable, all other must remain the same.

The probabilistic RMS packages usually employ Monte-Carlo simulations. Probability distributions are employed to represent the predicted occurrence of risks on the project mode. The model is then simulated interactively, each iteration generating a different value from each of the distributions employed. Frequency diagrams can then be developed from the consolidation of the results of the iterations performed to get a view of the likely completion cost of the project considered.

Where network based RMS is employed, the Monte-Carlo simulations are linked with network logic, which permits a more realistic representation of the project. The reason for this is that interdependencies between activities are represented and critical path calculations can be performed to identify the increased costs and durations of projects as critical paths change under the effect of identifies risks.

7.8 Classification of RMS

The classifications of RMS employed reflect the analytical methods and data management systems that the RMS packages utilise. The classifications considered are as follows:

- spreadsheet packages;
- network based packages;
- network and relational database packages;
- hybrid packages.

There is an inverse relationship between the complexity of a RMS package and its ease of use. Thus the spreadsheet packages, the least analytically complex of those considered, can be learnt in a matter of a few hours. However, the hybrid, network and relational database packages can take many weeks to master.

Spreadsheet packages employ Monte-Carlo simulation techniques to calculate the costs of undertaken projects or making decisions that are subject to uncertainty. There is no means of introducing networks, so these packages are principally of use where the decision alternatives are a number of well defined, discrete courses of action.

Network based packages permit the inclusion of interdependencies and precedence's between activities, so that these packages are well suited to the modelling of projects rather than just decision processes. Most network-based RMS packages allow the use of a number of time and cost performance parameters.

RMS employing networks and relational databases are usually bolt-on additions to planning packages. The large volume of data that these packages employ necessitates the use of databases if they are to run on networked personal computers. Batch processing facilities are employed to undertake the required computations.

The final type of RMS packages considered, the hybrid type, are tools for very specialised applications, typically employing complex statistical methods, for example, discontinuous logic mapping, data envelope analysis and fragility analysis.

The majority of the software packages commercially available include the same basic simulation options, either with or without networking.

Where the packages are based on networks, the network establishes the relationships between the work packages and activities, and calculates start and finish dates and float. The resources are defined, given a level of availability, and assigned to the activity network through the resource module. The cost module allocates costs to activities and resources, and keeps tract of the expenditure to monitor project performance. The risk module employs Monte-Carlo simulations to perform the required risk analysis, and the report module reads directly either from the activity networks or databases employed to produce either customised reports or data that can be exported to other packages to produce graphical representations of the outputs.

7.9 Selection of RMS

RMS packages can be effective tools in achieving project objectives if the packages employed match the project manager's needs and the project characteristics. Using the wrong RMS package may create a desire to make the project fit the package rather than vice versa. There are many different project management programs on the market with various functionality levels, and with prices that do not always match their functionality. Consequently, it can be hard to find a program that satisfies the project managers' requirements and hence evaluation guidelines are needed.

The most important thing to know when investigating the use of RMS is what you are going to do with it. If the required application and the associated computer functions needed to support the application can be visualised, the choice of suitable RMS will be narrowed to just a few packages. The selection of the most appropriate package is then much easier.

In terms of assessing individual RMS packages, there are two major features to consider. The first is the user-friendliness of the package – the time required to learn the package, to build models and perform simulations. The second is the results sought and produced – whether the package produces a result close to the result of the real life project that it is supposed to reflect. One can accept discrepancies between the model and the project if the model is user-friendly and flexible, while an accurate result is expected if the package is time consuming to use.

A decision table approach is outlined which uses multi-criteria decision making to determine the most suitable RMS packages for particular applications. The potential criteria against which RMS can be appraised include user-friendliness, modelling flexibility, screen and report graphics, heuristics, customising options, their ability to communicate with other programs and their purchase and maintenance costs and the cost of training to bring analysts up to an acceptable level of competence.

The contents of the decision table should give room for both subjectivity and objectivity in the information. The results obtained from the selection system are guidelines for further action, and cannot be used as scientific proof. A typical decision table is outlined in Figure 7.2. The super-factors

Super-factor	Super-factor A	Super-factor B
Super-weight	1–5	1–5
Sub-factor	Sub-factors	Sub-factors
Sub-weight	1–5	1–5
Program A — Sub-rate A B	1–5 Sub-weight × sub-rate $\Sigma A/\Sigma$ Sub-weight	1–5 Sub-weight × sub-rate $\Sigma A/\Sigma$ Sub-weight
Program A — Program rate	$C = \Sigma (B \times \text{Super-weight})$	
Program B — Sub-rate A B	1–5 Sub-weight × sub-rate $\Sigma A/\Sigma$ Sub-weight	1–5 Sub-weight × sub-rate $\Sigma A/\Sigma$ Sub-weight
Program B — Program rate	$C = \Sigma (B \times \text{Super-weight})$	

Figure 7.2 The basic appraisal system.

and sub-factors are the decision criteria against which the selection of the RMS package is made. The super-factors are the most important selection criteria for the user, and are given the weights between one and five. The super-factors are then divided into sub-factors, and the sub-factors are given sub-weights between one and five based on the analysts' priorities.

The RMS packages are then given a sub-factor rate between one and five based on how well they meet the sub-factor criteria. The subweights and the sub-factor rates are then calculated and added up, and this sum is divided by the sum of the subweights in order to find a mean rate, B. This mean rate is then calculated with the super-weight and a rate for the super-factor is found. The program rate, C, is found by adding the rates for the super-factors. The program with the highest program rate should be the most feasible for the user.

Let us now consider the following example of the selection process for a RMS package. It is decided by the analyst that, for a given application, the RMS packages should be appraised based on the appropriateness of program (AoP) and cost, these being the two super-factors employed. AoP is divided into the following sub-factors – start up, ease of use, network, resource, cost modelling, updating, reporting, risk and export/import. The cost super-factors are given the values five and two, respectively, reflecting the priority of the ability of RMS to model the project considered effectively over the cost of the RMS.

Figure 7.3 provides the decision table for the selection process, the sub-factor values and program values as indicated. In this case, the table is based upon the packages' ability to model packages at the earliest

	Super-factor	Appropriateness of program									Cost		
	Super-weight	5									2		
	Sub-factor	Start-up	Use	Network	Resource	Cost	Updating	Reporting	Risk	Export/import	Buy	Learn	Maintain
	Sub-weight	2	4	5	4	4	3	5	3	3	5	3	3
Open	Sub-rate	4	4	5	3	2	5	4	3	3	3	2	2
	A	8	16	25	12	8	15	20	9	9	15	6	6
	B	122/33 = 3.7									27/11 = 2.5		
Plan	Program rate	23.5											
SuperPro	Sub-rate	4	4	3	2	2	2	3	1	2	4	3	2
	A	8	16	15	8	8	6	15	3	6	20	9	6
	B	85/33 = 2.6									35/11 = 3.2		
Expert	Program rate	19.4											
CASPAR	Sub-rate	3	3	3	2	5	0	2	5	1	4	2	2
	A	6	12	15	8	20	0	10	15	3	20	6	6
	B	89/33 = 2.7									35/11 = 2.9		
	Program rate	19.3											

Figure 7.3 A general appraisal of the programs.

appraisal stage. From this example, it is evident that the third RMS package is the most appropriate of those considered. However, it should be noted that the choice made is a compromise, and that the package chosen is not the most appropriate in all aspects.

7.10 Modelling a project for risk management

Most RMS packages required a form of project model as a basis for understanding risk analysis.

Advantages

There are many advantages of using a computer to model a project; listed below are some of the more significant advantages:

❑ *Flexibility.* Computers are very flexible in the way in which they can accept information, enabling most projects to be modelled using a computer.
❑ *Speed and accuracy.* When the complexity of a model is such that no manual analytical technique can be used, computers often provide the

only means available for modelling. A computer can carry out many complex calculations very fast, compared to humans, and reliance can be placed on the accuracy of the calculations.

❑ *Additional reality*. Computer simulation enables real life complications, such as exchange rates, inflation rates and interest rates, to be included in the project model, and to calculate their effect on the project economic parameters.

❑ *Assistance in the decision-making process*. The project model enables a number of *what if* question and possible scenarios to be simulated, and shows the effects in terms of the outcome of the project. This simulation process shows the way in which the project is expected to react to certain events or changes and allows contingency plans to be drawn up that can be used in the event that any of the scenarios occur.

❑ *Scenario analysis*. Often there is no historical data available that relates to the project scenarios drawn up by the project organisers, so computer simulation is the only way to see how the project might react.

❑ *Reduces the dependence on raw judgement*. Few people have a reliable intuitive understanding of business risk, and risk modelling using computers removes the reliance on this intuition.

Limitations

There are a number of limitations to using a computer to model a project; listed below are some of the more significant limitations:

❑ *Poor data leads to inaccurate model*. A model of a project is only as good as the data that is input.

❑ *Model not representative of actual project*. Even if the data is accurate, it is possible that an inexperienced modeller can create a model that is not representative of the actual project. It is necessary for the project modeller to have a thorough understanding of the particular project to be modelled and of the RMS package being utilised.

❑ *Too easy to create inaccurate models*. RMS programs are designed to be user-friendly, which increases the dangers associated with the inexperienced/novice modeller.

❑ *Heavy reliance on subjective judgement*. Data is not always available when the project is being modelled, and some assumptions have to be made in order to complete the model. Often, a heavy reliance is placed on subjective information and personal judgement.

❑ *Not able to fully reflect real life complications*. The model produced is only a mathematical representation of real life and, therefore, does not necessarily accurately reflect the reaction of the actual project to real life complications. It is impossible to be sure that the model will react in exactly the same way as the real project because the project has not occurred. However, the more realistic the project model, more likely that the model will behave in a similar manner to the project.

❏ *Reliance on computer output.* Too much reliance is placed on the output from computers. It is difficult to tell whether a project model is an accurate representation of reality or not. If the model is very inaccurate, this can be easily detected. It is in situations where the model is almost, but not exactly, accurate that problems arise, because the model does not react to real life complications in the same way that the actual project would.

7.11 Data requirements for realistic modelling

The most important factor in the modelling of projects is the correct use of realistic data. However, in order to make a project model as realistic as possible it is necessary to include real life factors, such as, inflation rates and interest rates. The people who appraise a project need to know the implications of these factors, and understand the extent to which changes in these factors will affect the viability of the project.

The successful realisation of a project will depend greatly on careful and continuous planning. In most cases, input from a project manager will form the basis of the plan. Sequences of activities will be defined and linked on a timescale to ensure that priorities are identified and that efficient use is made of expensive and/or scarce resources.

Remember, however, that because of uncertainty it should be expected that a plan would change. The plan must therefore be updated quickly and regularly if it is to remain a guide to the most efficient way of completing the project. The software program should therefore be simple so that updating is straightforward and does not demand the feedback of large mounts of data from busy men – and flexible, so that all alternative courses of action are obvious.

RMS and project management packages use specific technology and some of the key terms are defined below:

❏ *Resources* – the organisation and utilisation of *Labour, plant*, and *Materials.*
❏ *Activities* – packages of work that consume resources and are defined by consideration of:
- the type of work (and therefore the type of resources required);
- location of the work;
- any restraints on the continuity of the activity.

❑ *Logic* – the relationship and interdependence between activities.
❑ *Duration* of each activity. This depends on:
  ▪ the level of resources allocated to the activity;
  ▪ the output of those resources;
  ▪ the quantity of work to be completed within the activity.
❑ *External restraints*, such as the specified completion date for the whole or some part of the work, the delivery date for specific material or restrictions on access to parts of the works.
❑ *Total demand for resources* accumulated from the individual activities.
❑ *Future problems* – potential difficulties and critical activities must be identified.

Factors that may be included in a model to make it more realistic include the exchange rate, inflation rate, interest rate and discount rate. These factors introduce non-project risks into a project, but it is important to be aware of these risks and to have an understanding of the impact that a change in any of these factors would have on the outcome of the project.

Inflation rate

Every country has price inflation to some degree and, although in some countries the inflationary problem is more acute than in others, this remains a problem that cannot be ignored. The inflation rate can be a significant factor in the success or failure of a project, particularly in projects of protracted duration. This factor is one that should be included in the modelling of every project, but those modelling the project must be careful when choosing the level(s) of inflation to be included. All goods and services are subject to different levels of price inflation, and so the overall rate of inflation applicable for a particular project is likely to be different to the average rate of inflation.

The average rate of inflation can vary notably over a few years, and is affected by a change of government policy. When modelling a project undertaken in the United Kingdom with duration of greater than five years, predicting the inflation rate can prove difficult.

Interest rate

The cost of engineering projects is increasing and companies have to borrow larger sums of money from the banks to finance these projects. The rate of interest payable to the banks on this money borrowed can appreciably increase the cost of a project. This is particularly the case in the public sector, private finance initiative (PFI), projects where the payback period often does not occur until a number of years after the start of the project. In a large number of projects the interest rate can make the difference between a project being viable or being loss-making.

The banks, that lend capital for the completion of engineering projects, create their own models to determine whether projects are viable. Interest rates vary over time, as a result of the economic climate, and this is another factor that can change with a change of government policies.

Discount rate

The discount rate is a factor that is utilised in net present value (NPV) calculations. The discount rate is applied to project cash flow figures to bring these values back to their present value. By discounting the cash flows, the financial benefits of all the projects are reduced to like units. This enables a number of different projects with different cash flow patterns and time spans to be compared, in terms of the value to the organisation undertaking the project. The discount rate(s) used for any given project is dependent on several factors, and determined by those assessing the feasibility of a project. The minimum level of the discount rate should be representative of the inflation rate(s) predicted for the duration of the project. However, the discount rate can be set at a level that takes into account the interest rate for capital borrowed. It is up to those who are modelling the project to decide what they wish the discount rate to reflect.

As with the inflation rate, the discount rate(s) for projects of long duration are difficult to predict with accuracy. The figure(s) chosen should be based on past trends and future predictions. In most cases, a discount rate based on historical data and future predictions is better than including no discount rate into a project model, particularly when the outcomes of several project options are to be compared.

Exchange rate

The exchange rate fluctuates and this causes difficulty in predicting an exchange rate that is suitable for application to a project model over a given time period.

7.12 Choice of variable distribution

When using a probabilistic risk analysis technique it is necessary to choose a probability distribution that is representative of the range and way in which the values of a variable might vary. There are a number of different types of probability distribution that can be chosen to represent the probability of occurrence of a range of values that comprise a variable. The shape of the probability distribution is often chosen based on historical data, which would show the distribution of the outcomes for this variable on past projects. The use of historical data is an objective means of deciding the shape of the probability distribution for a variable. However, historical data from which to choose the shape of a probability distribution is not always available. In the event of there being no historical data available it is left to the analyst and those associated with the project to decide on the probability distribution shape.

There are four commonly presented probability distributions and they are – the normal, beta, rectangular and triangular distributions. However, in practice not all distributions are symmetrical. It is often the case that probability distributions are skewed because it is more likely, for example, that there will be a delay in completing an activity than the activity being completed early.

It is important to choose a probability distribution that is appropriate for the variable being considered. If there is no historical data available then careful consideration should be given to the variable and its likely result, before a decision is made about the shape of the distribution. Utilisation of the experience of members of the organisation, from previous similar projects, might assist in the decision as to which probability distribution is most appropriate for a particular variable.

7.13 Case study

For the purposes of verification, the data and variables to be input to create the project models were decided upon after structured interviews had been carried out with industrial experts. The results produced from these computer models have been examined, in consultation with industrial experts, over the full range of the viable project domain, and in all cases, sensible results have been obtained. However, if any irrationality in the results had been discovered then further investigation would have been required.

The new industrial plant project is a hypothetical, but realistic, case study originally developed by Professor P.A. Thompson at University of Manchester Institute of Science and Technology (UMIST). The case study concerns the design, construction and operation of an industrial plant. It considers all stages of the project, from the start of the feasibility study through to the end of operating life of the project. The main objective of this project is to complete the construction phase on time, so that production output can reach the market within the desired *market window*. Although the cost of the project is a consideration it is not the main concern of the client/project manager, and a small increase in the cost of the project may be tolerated in order to ensure the timely completion of the construction.

The new industrial plant is estimated to have a pre-contribution size of 0.5 years (6 months), construction of the plant taking 2.5 years (30 months) to complete and the plant, when completed, is operated for 10 years (120 months).

Completion at month 37, is necessary because the client wishes the production output to reach the market place before any competitors can, and in that way the demand for the produce should reach the forecast levels.

The new industrial plant is to be built in two stages. The first stage of the project will have a maximum production output capacity of 5000 units per month. The second stage of the project will only be constructed if the demand for the production output during the first stage reaches or exceeds the levels of demand that were forecast.

Modelling

This project has been divided into 20 activities, which represent the main activities necessary to complete the project. Of these 20 activities, 17 are related to the construction of the project and the remaining three are related to the operation of the plant, Table 7.1.

The feasibility study is the start activity in this model, which is followed by the consideration of the report produced during feasibility study. If the report is accepted then the project will be sanctioned and then work can begin on the detailed design of the plant and an application for planning permission can be submitted. The design phase of the new industrial plant project is divided into four activities, each relating to the design of a particular section of the plant. The planning permission activity takes place in parallel with the detailed design activities because the outline designs for the project were completed during the feasibility study activity. The land purchase activity overlaps the planning permission activity by one month, allowing a suitable site to be chosen for the project by the time that the planning permission have been agreed. The purchase of the land must be completed before any construction work can take place.

The construction of the plant is broken down into five activities, one activity relating to each of the design activities and another activity for the construction of the services required for the plant. The services activity is a construction activity with no corresponding design activity because it relates to the construction of the site services, such as entrance roads

Table 7.1 New industrial plant – activities.

Activity number	Activity name	Duration	Start month	Finish month	Start date	Finish date	Total float	KNGS	D.MEN
				Timing of activities – earliest start					
1	Feasibility study	5	1	5	1995.01	1995.05	1	2	
2	Consider report	1	6	6	1995.06	1995.06	1		4
3	Design loading bay	5	7	11	1995.07	1995.11	19	2	10
4	Design office and labs	9	7	15	1995.07	1996.03	5	2	
5	Planning permission	3	7	9	1995.07	1995.09	1		4
6	Design rank farm	3	7	9	1995.07	1995.09	18	2	12
7	Design process plant	12	7	18	1995.07	1996.06	3	4	
8	Land purchase	6	9	14	1995.09	1996.02	1		
9	Construct loading bay	6	15	20	1996.03	1996.08	16	2	
10	Construct offices and labs	18	15	32	1996.03	1997.08	4	2	
11	Services	6	21	26	1996.09	1997.02	7	2	
12	Construct tank farm	6	21	26	1996.09	1997.02	7	2	2
13	Construct process plant	18	15	32	1996.03	1997.08	1	8	
14	Commission stage 1	4	33	37	1997.09	1997.12	1	2	
15	Operate stage 1	61	37	97	1998.01	2003.01	0		
16	Design stage 2	6	80	85	2001.08	2002.01	0		
17	Construct stage 2	12	83	94	2001.11	2002.10	0		

18	Commission stage 2	3	95	97	2002.11	2003.01	0
19	Operate stage 2	35	98	132	2003.02	2005.12	0
20	Operate stage 3	24	133	156	2006.01	2007.12	0

Hammock number	Hammock name					
			Timing of hammocks – earliest start			
1	Ouput of product/tonnes	120	37	156	1998.01	2007.12
2	Engineer/man months	35	1	35	1995.01	1997.11
3	D. men/man months	29	7	35	1995.07	1997.11
4	Admin. expenses	21	15	35	1996.03	1997.11
5	Energy	120	37	156	1998.01	2007.12
6	Engineer/man months	18	80	97	2001.08	2003.01
7	D. men/man months	6	80	85	2001.08	2002.01
8	Admin. expenses	15	83	97	2001.11	2003.01
9	Input A/thousand tonnes	120	37	156	1998.01	2007.12
10	Input B/thousand tonnes	120	37	156	1998.01	2007.12

Note: New industrial plant: project start date 1995.01: project finish date 2007.12. The start date refers to the start of the numbered month and the finish date to the end of the month.

and facilities for the workers. The construction of the offices overlaps the corresponding design activity by two months and the construction of the process plant starts six months before completion of the design of the process plant. These overlaps of activity can occur because, in some cases, sufficient design work has been completed to enable construction to start before the design activity has been completed. In this project, these activities have been overlapped to ensure that the plant can start operating exactly three years after the start of the feasibility study. Preceding the commissioning activity is three of the five construction activities, construction of the tank farm, services and construction of the process plant. This is because each of these sections requires testing before the plant can start operating. The loading bay and offices only require checking prior to the start of operation, as there is no equipment installed in these sections that requires testing, so they directly precede the first operating activity. If there are no problems encountered during the commissioning activity then the plant can start operating as planned. In the first month of operation, the plant produces 2000 units and this steadily rises until at the end of the activity the plant is producing 5000 units per month. The duration of this activity is dependent on the level of production output and the growth in demand. Assuming that the demand pattern is as forecast, this activity has duration of 61 months.

In the second stage of the project, the construction of the extension to the plant, starts 21 months before the end of the first operating activity, provided that demand during the first operating activity reaches the required level. The first operating activity has been operating for 40 months before the start of stage two, and this is thought to be sufficient time to decide whether demand for the production output is as forecast or not. Stage two of the project comprises of three activities, design, construction and commissioning, respectively and each of these activities follows the previous one without any overlaps. The second operating activity has duration of 35 months and during this time, the production output is expected to rise from 5000 units per month to the new maximum production capacity of 8500 units per month. The third operating activity has duration of 24 months, and during this activity, the production output is slowly decreased, at the rate of 5% per month. This represents the growing inefficiency of the plant as it comes to the end of its working life, the outdated technology and the increasing number of mechanical failures that would occur.

The key resources used in the new industrial plant project are modelled as the cost of the draughtsmen and engineers, the administration costs of the project, the raw materials required in the production process, the cost of the energy used in the production process and the production output.

Each of the resource costs input into this model was related to the length of time that the resource was used. In this model, the sales price per unit of production output has been taken to be £100.

In the project model, a number of cost centres were defined and all costs and revenues input into the model were attributed to these cost centres. The cost centres were divided into construction cost centres and operational cost and revenue centres. The construction cost centres for the new industrial plant project are technical costs, plant costs, construction costs and capital costs. The operational cost centres for this model are raw material costs, direct costs, indirect costs, overheads and revenue. The majority of the costs included in the model are mainly input as activity – time related costs, with a few costs such as the cost of purchasing the land, being included as activity fixed costs.

In creating this model, it was necessary to make some assumptions about the project and the way in which it should react in certain situations. An initial assumption included in the new industrial plant model is that the two stages of the project goes ahead as predicted in the feasibility study, and that demand for the production output is great enough to sustain this assumption. This assumption gives the base case model. It is also assumed that the revenue per unit of the production output remains unchanged throughout the operating life of the project.

Risk variables

Using the principles of risk identification, a large number of risks were identified and a sensitivity analysis was carried out. Only those risks having a high input and judged to have a penalty of revenue were included in the project model. Risks in the new industrial plant project have been considered at two stages, at the sanction and the commissioning stages of the project. The spectrum of risks will have changed considerably between these two points in the project. This is because at the sanction stage, the client has to consider all the possible risks in the project, but at the commissioning stage the risks associated with the construction of the plant are no longer relevant and the client is then only concerned with the risks associated with operating the plant.

Computer software using an activity network was adapted for the modelling of this project. Nine risks were modelled for the sanction stage and four risks modelled for the commissioning stage of the project. The risks included in the model at the sanction stage of the project were:

❏ delayed planning permission;
❏ delay in the design activities;

❑ delay in the purchase of the land;
❑ delay in the construction of the plant;
❑ change in demand for the production output;
❑ change in the cost of raw materials;
❑ change in the cost of energy;
❑ change in the revenue from the production output;
❑ change in the engineering costs of the project.

The risks included in the model at the commissioning stage of the project were:

❑ change in demand for the production output;
❑ change in the cost of raw materials;
❑ change in the cost of energy;
❑ change in the revenue from the production output.

The computer program contains limits on input data. To satisfy the limitation the data was rounded up to the nearest sensible value. The timing option used for each of these case studies was months, and therefore the output from the models can only be considered in terms of a minimum of one month's cost. The output from the models used in this chapter has been adjusted to reflect this level of accuracy.

This model was used to calculate the economic parameters for the project, which provide a basis for comparison between the various simulations that were carried out on this and other project models. The economic parameters relating to this project model, shown in Table 7.2, were derived by the computer from a deterministic analysis of the project.

The new industrial plant project model is now ready for simulating analysis and hence assessing the likely outcome of the project.

Table 7.2 New industrial plant – economic parameters.

IRR	37%
Present value	£88 040 000
Payback period	6 years
Maximum investment	37 450 000 (£ years)
Return	215 700 000 (£ years)
Maximum cash lock-up	£13 720 000

Note: No discount rate and no inflation rate applied to the figures.

7.14 Case study simulations

A number of simulations were carried out using this project model and each of these simulations has been documented separately. Some of the simulations were designed to introduce non-project risks into the project model and the remaining simulations were carried out to determine the sensitivity of certain risks on the outcome of the project.

Discount rate

The discount rate affects a number of the project economic parameters, such as the NPV, the payback period, the net return, the cash lock-up and the maximum investment. In the base case model, no discount rate was applied to the cash flow figures.

A discount rate of 6% was imposed on all the project costs and revenues. The figure of 6% was chosen because it was felt to be marginally higher than the rate of inflation at the time of writing this book and would be likely to represent an average rate of inflation when taking into account the duration of the project.

Inflation rate

A single rate of inflation of 5% was chosen for each of the cost centres and this remained constant for the duration of the project. This rate was chosen because it is similar to the rate of inflation and is a realistic rate for the first few years of the project. This rate might not be as applicable after the first five years of the project, but in the absence of an indication of the trend for the inflation rate over the duration of the project it was felt that a constant rate of inflation should be applied.

Finance charge

A finance charge of 12% was imposed on total yearly cash outflows until the project reached payback. The level of the charge was set at 12% because that is the rate at which banks were currently lending to businesses. This charge may not be applicable for the duration of the project, but due to the influence of external factors, it is very difficult to predict with certainty the level of this charge in the future. The yearly cost of this charge was modelled by introducing a new cost centre and placing the cost, as a fixed cost at each year end. This is a crude method for costing and allocating a finance charge, however, due to the uncertainties surrounding the appropriate level to be charged this was felt to be appropriate.

Sensitivity analysis

The reason for carrying out a sensitivity analysis was to identify areas of the project that require particular attention or areas that require a more detailed study before work begins.

Due to the dynamic nature of the risks, the sensitivity analysis was carried out in two stages, at the sanction and commissioning of the project. In order to analyse the sensitivity of the project at these two stages it was necessary to create two computer models of the project, identical in all aspects except for the key risks, or variables, being modelled.

The ranges of variation that were applied to these risks have been detailed and were chosen because they represented a range within which the variable might reasonably be expected to deviate. These ranges were not chosen based on historical data because this information was not available, however, they were chosen with assistance from experienced project personnel. In a sensitivity analysis the ranges applied to the variables are not as critical as in a probability analysis because the technique is used to consider the effect of change in the key risks on the economic parameters of a project and gives no indication as to the likelihood or occurrence of the variables analysed.

Probability analysis

A probability analysis considers the key variables in combination, unlike the sensitivity analysis which considers them in isolation. The probabilistic analysis technique used on the new industrial plant project model was the Monte-Carlo technique. The analyses carried out on the model were run for one thousand iterations, in an attempt to reduce the statistical biases inherent in this sampling technique. The output from this analysis gave a range of estimates for the outcome of the project, ranging from the most pessimistic to the most optimistic outcome.

The probabilistic analysis of this project was also carried out in two stages. None of the key variables included in the project models were correlated despite some indications for links between some of the variables. The reason for this was that the amount of correlation between these variables was unknown.

Sanction risks

The ranges chosen for the variables relating to the delay in design and the delay in construction only depict the possibility of a delay. This was because the estimates of the duration of these activities represent a target completion time, and although this time was thought to be achievable it is not expected that the work will be completed any earlier than this. A triangular distribution was applied to both of these variables because it was felt that they could be predicted with some confidence, and that the estimates used in the model were slightly optimistic. The variables relating to a delay in planning permission and a delay in the purchase of the land were given wide ranges of variation and uniform distributions. The reason for this was that both of these risk variables could either pass with no problems encountered or there could be long delays in both processes. These variables could not be predicted with any certainty; therefore, a uniform distribution was applied to both of the variables. The engineering costs variable related to the cost of construction of the process plant activity, which was the most costly construction activity. The cost of this activity was also related to the time taken to complete the activity, so any difficulties encountered or any delay in this activity would affect the costs. In this case it was likely that the costs of the activity would rise rather than fall, although allowance was made for the possibility of a small decrease in these costs. It was felt that the costs could be predicted with some confidence so a triangular distribution was applied to this variable.

The energy costs for the project were given a wide range of variation, with the implication that the cost of energy would increase rather than decrease. A uniform distribution was applied to this variable, which indicates that this variable cannot be predicted with confidence. The cost of the raw materials required to make the product is an important factor in the operating costs of the plant, however, the actual raw materials is known within a fairly small range of variation, hence a triangular distribution was applied. The variables relating to the demand for the product and the revenue from the product were given wide ranges because of the uncertainty over the market acceptance of the product. A triangular distribution was applied to the variable relating to the revenue from the product. However, a uniform distribution was applied to the demand for the product, because at this stage it was difficult to predict with any certainty how consumer tastes would vary over time.

Commissioning risks

The key variables included in this project model were based on the variables included in the sanction stage model. The ranges of variation applied to these variables were reduced from those applied in the sanction stage model because at this stage in the project there was less uncertainty about the market. The information that was available could be used immediately to predict the outcome of the project. At this stage all the variables

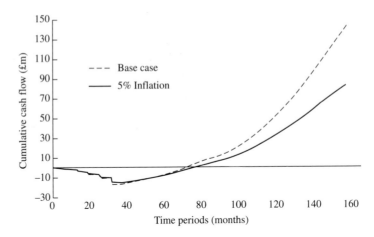

Figure 7.4 New industrial plant – cumulative cash flow.

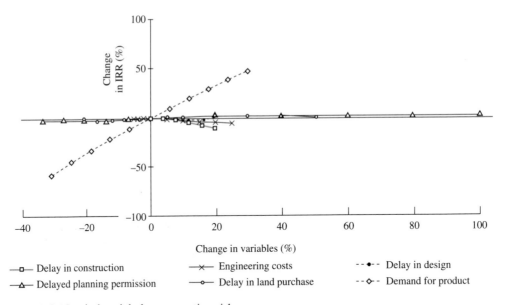

Figure 7.5 New industrial plant – sanction risks.

were given triangular distributions because they could be predicted with some confidence.

7.15 Analysis of the result

The results of these simulations are shown in Figures 7.4–7.7 and Table 7.3 in terms of their effect on the project economic parameters.

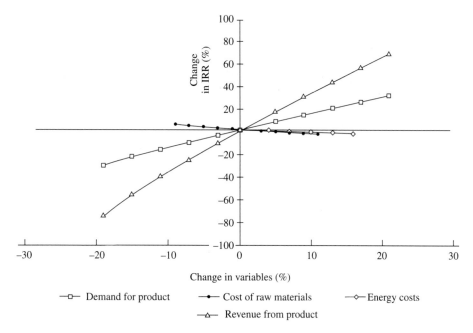

Figure 7.6 New industrial plant – commissioning risks.

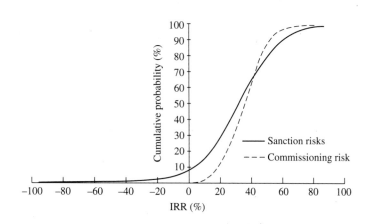

Figure 7.7 New industrial plant – cumulative frequency plot.

Discount rate

The application of a discount rate to the project cash flow has the effect of decreasing the importance of cash inflows near the end of the project and increasing the importance of cash outflows at the beginning of the project. This can be seen when the results of this simulation and those of the base case model are compared, as in Table 7.3. When the discount

Table 7.3 New industrial plant – data.

Description	Base case	Discount rate 6%	Inflation rate 5%	Finance charge 12%
IRR (%)	37	37	44	31
(£)				
Present Value	88 040 000		147 010 000	83 010 000
NPV		47 290 000		
(Years)				
Payback	6		5.5	6.5
Discounted payback		6.5		
(Years)				
Maximum investment	37 450 000		38 820 000	51 990 000
Discounted maximum investment		36 170 000		
(Years)				
Return	215 700 000		369 400 000	167 490 000
Net return		102 680 000		
(£)				
Maximum lock-up discounted	13 720 000		15 050 000	16 330 000
Maximum lock-up		12 290 000		

rate was applied the NPV was almost halved, as was the net return, and the payback period was increased by about 6 months. These economic parameters are, largely, determined by the level of the cash inflows near the end of the project. The maximum investment and maximum cash lock-up, which represent the amount of money tied up in the project, were both slightly reduced.

Inflation rate

Figure 7.4 shows the cumulative cash flow for the base case project model and the cumulative cash flow with a 5% inflation rate applied to each of the cost centres. From the diagram it can be seen that the payback period for the project was decreased by about 6 months, the return and the present value (PV) were both dramatically increased and the IRR was also increased. The maximum investment and maximum cash lock-up were both slightly increased, although not to the same extent as the return and PV. Figure 7.4 shows that from the time the project reaches payback to the end of the project, the cash inflows are exceeding the case outflows by an ever-increasing amount. This was in keeping with what would be expected since the operating costs for the project were significantly less than the revenue generated, so when an inflation rate was applied the difference was exacerbated.

Finance charge

The finance charge is an amount paid on the project cash outflows before payback is reached. Figure 7.4 shows the cumulative cash flows for the base case project and the cumulative cash flow for the project with the inclusion of a finance charge. From this diagram it can be seen that if the maximum investment required for the project rose by 5% inflation then the payback period increased by about 6 months.

Sensitivity analyses

This was carried out in two stages, for this study. The results of these analyses are shown in Figures 7.5 and 7.6. These spider diagrams represent the change in the key variables in relation to the change in the IRR for the project. The reason for considering the change in the IRR, as opposed to the PV or any other economic parameter, was that the IRR is the most applicable economic parameter for projects with a total duration of more than 5 years.

The sensitivity of the project to these variables is shown by the gradient of the lines. In Figures 7.5 and 7.6 the lines that tend towards the vertical axis are the variables to which the project is most sensitive and those lines that tend towards the horizontal axis are the variables to which the project is least sensitive to. Any variable producing a percentage change in the IRR of greater than the percentage change in the variable is regarded as a variable to which the project is sensitive.

Sanction risks

It can be seen, from Figure 7.5, that the variable the project was most sensitive to is change in the demand for the product. The line representing the change in the revenue from the product has not been fully included in the diagram since a change of more than −20% in this variables gives rise to a negative IRR, and there is no purpose in showing that on the diagram. A small change in either of these variables has a very large impact on the project, in terms of the project IRR.

These two variables dominate all the other variables plotted on the diagram; however, the other variables must not be ignored. The line representing a delay in construction is also quite prominent, from a +12%

change onwards the lines moves towards the vertical axis. On a diagram with a smaller vertical scale this change would be much more noticeable. The reason for this change is because, by this stage, the float in the programme has been used and any further delay in this variable is central and delays the start of the operating stage.

The variables representing a delay in the design, delayed planning permission and a delay in the purchase of the land all appear to have little effect on the IRR. The probable cause of this is that these activities are of a relatively short duration, in comparison with the construction or operating activities. A change of +25% in the duration of a 4 month activity only increases the duration of the activity to 5 months, whereas for an activity with a duration of 16 months a +25% change increases the activity duration to 20 months. Therefore, despite the large percentage changes in these variables, it is probable that the float in the programme has not been completely used and, therefore, a delay in these variables would have no impact on the start of the operating activities.

The inverse relationship between the energy costs and the IRR is one that would be expected. This variable shows that any small increases in the cost of energy occurring during the operation of the plant will decrease the project IRR. The variable representing the raw material costs for the project follows a similar pattern to the energy cost variable but the impact on the IRR is slightly greater in this case.

Commissioning risks

Figure 7.6 shows the spider diagram created from the results of this sensitivity analysis. The variables relating to changes in the cost of energy and in the cost of the raw materials follow a similar pattern to those in the sanction stage analysis. Both variables show an inverse relationship with the IRR and the lines are straight representing a constant level of change in the economic parameter. The project remains more sensitive to a change in the cost of the raw materials than it is to a change in the cost of energy.

The variables representing the demand for the product and the revenue from the product still dominate the diagram, meaning that the project is very sensitive to any changes in these parameters. The project is most sensitive to changes in revenue from sales of the product. A change of +10% in this variable leads to a change of about −40% in the IRR of the project. At a change of −16% in the revenue from the product it can be seen that there is a definite change in the gradient of the line. This suggests that at this point something happens in the project and a possible reason for this change in the gradient of the line could be that the operating costs reach equality with the revenue.

Probability analyses

This was carried out for sanctioning and for commissioning. The results of these analyses have been represented in the form of one cumulative frequency diagram, Figure 7.7. This diagram represents the range of values for the IRR that the project might achieve and shows the cumulative probability of each of those values being achieved.

The results of these analyses will be considered between the cumulative probability range of 15–85%. These limits represent approximately one standard deviation removed from either end of the frequency distribution. The information contained in the 30% of the diagram not considered represents the tail ends of the distribution. The frequency of occurrence of these values was very small and the range over which the values were spread was very large. The probability of achieving a particular percentage IRR for the project can be determined by deducting the percentage cumulative probability from 100.

Sanction risks

The deterministic estimate for the IRR of the project is 37%. However, from the probability analysis, it can be seen that the outcome of the project, in terms of the IRR, over the 15–85% range is from about 10% to 55%. Figure 7.7 illustrates this and shows that the probability of achieving a 37% IRR is approximately 40%. There is a 50% chance of achieving approximately a 30% IRR for the project and a 70% chance of achieving approximately a 22% IRR for the project. The range of possible values for IRR given by this analysis was very wide, probably because of the sensitivity of the project to the market risks.

Commissioning risks

The range of values recorded for the IRR from the probability analysis was smaller than that recorded at the sanction stage. The range of values for the IRR in this analysis was from approximately 22% to 50%. Figure 7.7 illustrates this and shows that the probability of achieving a 37% IRR is approximately 50%. There is a 30% chance of achieving approximately a 43% IRR for the project and a 70% chance of achieving approximately a 30% IRR for the project.

It would appear from the analysis of this project model that the deterministic estimate for the IRR of the project was slightly optimistic. The reason that the chances of achieving an IRR of 37% have increased because the number of variables considered in this analysis was less than at the sanction stage, and the ranges of variation applied to those variables was reduced.

7.16 Discussion of findings

The first three simulations carried out on this project model were deterministic analyses to consider the impact of non-project risks on the project. The inclusion of a discount rate of 6% in the project model had the general effect of reducing many of the economic parameters to almost half of their values in the base case model. The inflation rate had the effect of making the project look more desirable, by increasing the parameters that indicate the overall profitability. However, this was to be expected because of the diverging values of the operating costs and the product revenue. The results gained from the inclusion of a finance charge had the effect of making the project seem less desirable. The reason for this being that it reduced the value of the economic parameters that indicate the profitability of the project. Each of these simulations was considered in isolation, although for a real project it would be reasonable to expect some combination of these factors.

Both of the sensitivity analyses were dominated by the market risks, which are the demand for the production output and the revenue received from sales of the output. The other variable to note at the sanction stage was the possibility of the construction phase being delayed.

The probability analyses showed the deterministic output from the project model to be slightly optimistic. The range of possible values for the IRR in the sanction stage analysis was very wide, probably because of the sensitivity to the market risks. At the commissioning stage the range of possible values for the IRR had decreased significantly, thus raising doubts about achieving an IRR of 37%. This reduction in the range of values for the IRR would be expected as a result of the reduced number of variables, which were considered and the reduced ranges that were applied to these variables.

From the results of the simulations carried out on the project models it can be seen that the objective of timely completion is not as important as gaining information about the market or securing a buyer for the production output. If the market for the production output does not exist then it is not of any importance whether the construction of the plant is completed on time. These results show the dominance of the market on the profitability of the project.

7.17 Summary

The use of IT is significant in the presentation of risk management. Most software packages require a basic project model to which the effects of key values are applied.

The case study sensitivity analysis at the sanction stage shows that the project becomes sensitive to a delay in the construction once the float in the programme has been used, because any delay beyond this point has a direct impact on the start of operation of the plant. A delay in the designs for the plant have a relatively small effect on the outcome of the project, however, this variable was not linked to the construction process. If these two variables were linked, or correlated, a delay would show a much greater impact on the outcome of the project than either of these variables does independently. If there was a delay in the design reaching the construction contractor then the construction process would be delayed, which soon begin to have an impact on the start of operation. The client might wish to consider including a clause in both the design and construction contracts that would lead to the contractor being penalised if a delay occurred. Since this analysis was carried out before the project was undertaken, another option open to the client is to award just one contract for the design and construction of the plant. This would mean that there was only one contractor who would have more control over the process, which should reduce the possibility of a delay in either the design or the construction processes.

Both the sensitivity analyses show that the market risks are dominant in this project, compared with the other risks. The sensitivity of the project to these risks affects the results of the probability analyses and indicates that the project might achieve only a moderate rate of return. To reduce these risks the client could carry out a market survey to assess the possible market for the product, and reduce the uncertainty about the market at the sanction stage of the project.

Chapter 8

Risk Allocation in the Contracting and Procurement Cycle

A client has the ultimate responsibility for project management; they must define the parameters of the project, provide finance, make the key decisions and give approval and guidance. The contractor (supplier) provides a service for the client. These parties must work together if a project is to succeed but, through the proliferation of claims, clients and contractors have become further removed, the construction industry has suffered, projects cost more and clients look elsewhere to invest money.

In this chapter, the procurement cycle is examined to highlight the diversity of approaches to the allocation of risk in the supply chain through organisational methods, payment mechanisms or a combination of both of these. The appropriate procurement or contract strategy will only become apparent as the evaluation progresses from initial appraisal to full analysis, including consideration of potential areas for dispute because of known and unknown risks.

8.1 Typical contracting and procurement processes

With reference to Figure 8.1, the following is a description of the main stages of the contracting process. When the project is conceived, the client has to undertake an initial risk assessment and evaluate whether the risks associated with the project are high or low. From this, the client will be able to decide whether the risk exposure is acceptable and from risk analysis decide whether to proceed with the project. It may be that the risks associated with the project are high but by careful development of the project execution plan a number of risks can be eliminated, transferred or insured.

From the project execution strategy, covering all aspects of engineering, procurement and construction for the whole project (including health, safety and environment studies), a contracting strategy can be developed, taking account of available strategies, selection procedures and recent

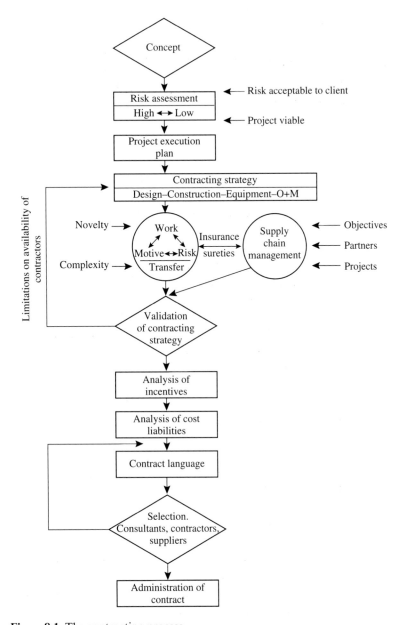

Figure 8.1 The contracting process.

trends. Projects should be mapped onto a chart, like that in Figure 8.2
to determine level of risk, control, motivation and design completion to
aid in the choice of strategy. The fundamental choices being made here
relate to work, motive and risk transfer through the supply chain, which
will be influenced by the novelty and complexity of the project, and by

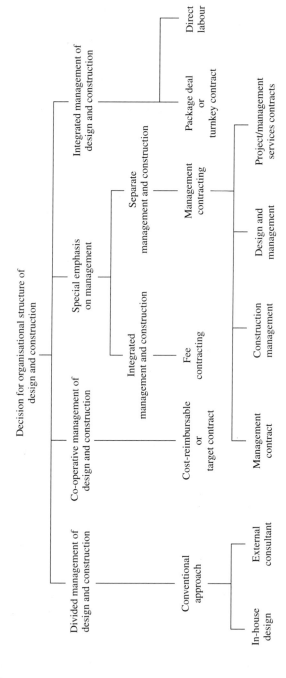

Figure 8.2 Choices available to project management for the organisational structure of design and construction.

other factors such as, whether the client wants to enter into partnering arrangements. The management of the whole supply chain is increasingly being recognised as crucial to the success of a business because it allows strategies to be developed that transfer risk to the party best able to manage that risk. For example, if concrete is a key component of the project then the concrete supplier should be brought into the project at an early stage when changes can still be made to design mixes and then risks associated with supply could be allocated to the concrete supplier. Insurance or sureties may be required where the value of damages (liquidated or non-liquidated) are large compared to the turn over of the supplier.

Possible strategies and reasons for their selection are illustrated in Figure 8.2 and will be discussed in detail later. Validation of the chosen strategy can then take place in view of the contractors available to carry out the work. Pre-qualification is key to this stage of the process.

The analysis of cost liabilities or impacts of risks and the contract language are the basis of risk evaluation and management. Risks cannot be eliminated through contracts but the strategy chosen for dealing with the risk can influence how they are managed and dictates how they are allocated. Selecting a contract strategy requires decisions about – type and location of project; number of packages; responsibilities of project team; flexibility required; project life cycle; terms of payment and the basis for selecting contractors.

The life cycle costs of any project should be considered in the value management exercise. Risks may be identified that could occur during the operation phase of the project or long after the construction has been completed. There is a temptation when carrying out the value management study to concentrate on design issues, equal importance should be given to identifying potential risks and the form of contract that best covers such risks, while still meeting other objectives, such as the management process and organisation costs.

Once the decision has been made as to the most appropriate contact strategy, the contractor can be selected, using a rigorous tender evaluation procedure. The contractual relationship is established during the tender period and during the early stages post contract award. If the project is to be a success then it is during this phase that alignment must be achieved. This could take the form of a series of workshops where any potential high risk sources are identified and joint plans are drawn up for dealing with them in the event that they materialise. The key objectives for the client are:

❏ to obtain a fair price for the work, bearing in mind the general state of the construction market at the time; and

❏   to enter into an agreement with a contractor who possesses the necessary technical skill, resources and financial backing to give the client
the best possible chance of the project being completed within the
required time, cost and quality standards.

Many clients are now entering into partnering arrangements to overcome
potential contracting problems. Partnering is a philosophy that aims to
bring together all members of the supply chain through mutual objectives
for the benefit of all parties. The idea being that as many of those involved
in the project as possible (throughout the supply chain) should develop
and sign up to a partnering charter, this charter spells out mutually aligned
objectives, including risk sharing, so that all parties are equally motivated.

Contract administration is the final stage of this process and it is here
that the selected strategy is tested. The occurrence of risk events and
its successful management will reflect upon the appropriateness of the
strategy selected. The risk of disputes is much greater in cases where an
inappropriate strategy has been forced upon the project team.

8.2 Value planning case study

This section illustrates how value management reviews can be utilised to
determine the most suitable contract strategy for a sewage treatment plant
(STP) procured to meet specific needs.

During the first review, the client has defined the following:

❏   the project is needed quickly to meet increased flows;
❏   the capacity of the plant is to be 50 000 m³;
❏   the favoured option of the plant is extended aeration with operation,
maintenance and training package options;
❏   the risks associated with the plant are performance guarantees, time
to construct, lack of client resources, design on clients and money
available.

The second review has defined the following:

❏   the processes to be provided to ensure low operating costs;
❏   a preferred outline design proposal and layout;
❏   the basis for continued design development as the objectives are met
such as future expansion;
❏   identification of the risks associated with the design chosen those
being, flexibility, availability of materials and meeting the required
performance.

The third review has defined the following:

❑ the selected design option meets all the objectives;
❑ the final design, specification and operation methods, schematics, hydraulic profile and probability impact (PI) diagrams;
❑ a proposal of how the project can be best implemented to meet the objectives and risk allocation.

In terms of contract strategy, it is clear that the following elements are required to meet the project objectives:

❑ early operation through fast-track construction methods;
❑ a two year operation, maintenance and training (OMT) provision;
❑ low operating costs.

The risks identified are those associated with:

❑ lack of client resources;
❑ no provision for design and guarantee of meeting the process;
❑ uncertainty regarding ground conditions;
❑ requirement to meet other works by a specified date;
❑ meeting the specification;
❑ meeting specific budgets;
❑ providing the most suitable technology;
❑ lack of knowledge with regard to disputes.

Questions that need to be asked are:

❑ What additional value is gained through the choice of contract strategy?
❑ What is the best contract strategy?
❑ Will the contract satisfy the needs of the project?
❑ Will the contract allocate risks in an equitable manner?
❑ What will the project cost under this contract strategy?
❑ What are the savings/additions for risk cover?
❑ Can the contract be implemented at the right price?

At this stage, the client seeks a contract strategy that will meet the project objectives and allocate the major technical risks to the contractor. In this case, a fixed price lump sum turnkey contract is identified as the most suitable contract strategy provided the project could be completed within the client's budget. The client can utilise model forms for the contract based on other projects procured in this manner.

The client can now perform a more rigorous risk analysis using qualitative and quantitative assessments. The key to this method is that many of the risks identified can be expressed in terms of the contract and risks allocated accordingly. Clearly, the contractor will accept many of the risks identified by the client at a premium. The client may quantify the cost of the premium for covering such risks as a percentage of the total contract price.

In this case, the client will bear the following risks:

- provision of the site and access;
- raising the finance;
- covering global risks such as political, legal and environmental risk.

The contractor will bear the risks associated with:

- process guarantee;
- project design;
- construction deadline;
- commissioning and then operating, maintaining and training personnel for a period of 2 years;
- the uncertainty of the ground conditions;
- providing a fully operational plant at a fixed price;
- commercial risks associated with the costs of materials and resources.

The life cycle costs of any project should be considered in the value management exercise. Risks may be identified that may occur during operation of the project, long after the construction has been completed. In this particular project value management techniques were used to determine the most suitable operator contract after the OMT contract was completed, in this case an affermage contract was deemed to provide the best value for the remaining life of the project.

8.3 Known and unknown risks in contracts

The three main functions of contracts are work transfer (to define the work that one party will do for the other), risk transfer (to define how the risks inherent in doing the work will be allocated between the parties) and motive transfer (to implant motives in the contractor that match those of the client). There is a basic conflict between these provisions and this chapter concentrates on optimising these provisions for a particular project.

The identification and allocation of risk is a lengthy process that will require a number of iterations for optimum results. During project appraisal, risks that could occur throughout the whole life of the project should be identified for the whole supply chain. These could then be considered based on:

❑   which party can best control events;
❑   which party can best manage risks;
❑   which party should carry the risk if it cannot be controlled;
❑   what is the cost of transferring the risk?

That is to say, some are pure risk, for example, force-majeure, while others are created, for example, by technology, by the form of contract or organisational structure. These are not the same. The client must ensure that through the contract strategy chosen his exposure to risk is optimised, considering both the up and down side.

Traditionally, risk in construction projects is allocated as follows:

❑   client to designer and contractor;
❑   contractor to subcontractor;
❑   client, designer, contractor and subcontractor to insurer;
❑   contractor and subcontractor to sureties or guarantors.

The impact of risk events on projects is, in the vast majority of cases, related either directly or indirectly to cost. Time delays inevitably have a consequential cost. Where materials, plant fail or the supplier of services does not perform, the additional cost is apparent. Where less tangible risk events occur, such as emissions or environmental disruption, no direct cost may be incurred immediately by the client but in these circumstances the costs may be incurred at a later date.

Client organisations should appreciate, when deciding upon the allocation of risks, that they will pay for those risks that are the responsibility of the contractor, as well as those that are their own. Contractors employ contingencies in their tenders as a means of guaranteeing their return in the event of construction risks occurring.

The payment mechanism employed, price or cost based, will determine the location of these contingencies. The allocation of risk between parties to a contract should be identified prior to tender. Tender documents illustrate the risks and responsibilities between the parties to the contract. In some cases, client organisations are now requesting potential tenderers to provide a risk statement as part of their tender. The risk statement provides the client with the risks often not covered in the contract that the tenderer feels may occur and how they would respond to such risks should they occur.

A number of clients now list potential risks in the tender documents and request tenderers to price each of them as part of the tender, the evaluation of such risks and the price for their cover being part of the tender assessment criteria. The size of the contingencies employed by the contracting parties will be dependent upon a number of factors which may include the following – the riskiness of the project; the extent of the contractor's exposure to risks; the ability of the contractor to manage and bear the consequences of these risks occurring; the level of contractor competition; and the client's perceptions of the risk/return trade-offs for transferring the risks to other parties.

When risk events that are the client's responsibility occur, the contractor should receive the funds necessary to overcome the particular risk event. Where there is some uncertainty over responsibility for a particular risk event, the contractor is entitled to pursue claims for additional payment from the client when it occurs. Clearly, the client is likely to wholly or partly pay for risk events irrespective of which party bears responsibility for them.

Contractors usually assess the cost or price of given risk events higher than clients. The reason for this is related to the long-term effects that risk events have on the business of the two organisations. This is particularly the case when a large client organisation employs small and medium sized contractors to construct small to medium sized projects. A cost overrun of 10% on a £1 million project would be a source of concern to a large client, but to a contractor in the current economic climate, with low margins, it could be the difference between staying in business and liquidation.

The risk-averse behaviour of contractors and risk neutral behaviour of major clients has been identified elsewhere. A risk neutral client is assumed to view a £1 loss in the same light as a £1 gain. For a contractor, the loss may be perceived as far higher. The effect of this is to make the contractor's estimate of the cost of a given project greater than that of the client if the responsibility for risk is evenly split between the two parties. Whilst this may not be reflected in the contractor's tender, it is likely to become apparent as the construction of the project proceeds and risk events occur.

Disputes are likely to occur between the client and the contractor when the risks are not clearly allocated. A number of authors have indicated the importance of good relations between the contracting parties to successfully complete projects. The recent trend towards partnering has been a response to the proliferation of claims. Partnering is a structured management approach to facilitate teamworking across contractual boundaries.

Its fundamental components are formalised mutual objectives, agreed problem resolution methods and an active search for continuous measurable improvements. It should not be confused with other good project management practice or with long-standing relationships, negotiated contracts or preferred supplier arrangements, all of which lack the structure and objective measures that must support a partnering relationship.

The critical success factor for partnering is the commitment of all partners at all levels to make the project a success. The result is that the partnering agreement drives the relationship between parties rather than the contract documents. When this is the case, the objectives of all parties are aligned, the whole supply chain is managed and mutually aligned objectives are sought.

8.4 Risk allocation strategies

For a particular project, one or several organisational structures may be chosen depending on the project's size, novelty and complexity. The design teams required (whether in-house or external); methods of management; supervision; certain restrictions such as political, social or economic; the available resources and expertise committed by the client must be considered as unique for a particular project and will affect the type of organisational structure to be chosen.

Risk allocation strategies should be determined at the inception of the project by the client. Figure 8.3 shows that effort at the start of the project in minimising the need for change will have maximum impact and effectiveness in realising a successful project. Further risk management, exercises may be undertaken during the course of a project but the reallocation of risk at this time is rare and will require negotiations with the contractor, which may or may not be successful. There will be

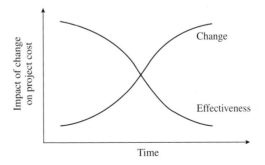

Figure 8.3 Relationships between effectiveness and cost of change over time.

occasions where no contractors will be prepared to bid for a contract that places arduous conditions upon them and the client has to reconsider the strategy. A contractor's exposure to risk must be related to the return that they can reasonably expect from a project. Thus, if a contractor is making only a 5% return on a project, it is reasonable for a contractor's risk exposure to be restricted. Alternatively, tenders may be much higher than expected reflecting the cost of transferring the risk to the contractor.

The main characteristics of the available choices of risk allocation strategy can be grouped according to organisational structure or payment mechanism, there is a tendency for certain payment mechanisms to be associated with certain organisational structures, where such relationships exist they will be highlighted and discussed. The choice of contract and hence risk allocation strategy is determined by the policy decisions of the client and the requirements of the individual project. On occasion, however, the policy considerations of the client take precedence with little regard to the project concerned. The client must remember that inappropriate strategy on the retention or distribution of risks will jeopardise the project.

Construction risks such as ground conditions, risk of non-completion, cost over runs and risk of delay are considered as major technical risks. Most construction risks are controllable and should be borne by the contractor whether the project is publicly or privately funded. Similarly, risks associated with labour, plant, equipment and materials, technology and management are controllable risks and should lie with the contractor(s).

Specification risk and errors in design that could have a detrimental effect on both construction and operation are also common. Physical hazards that may occur in the construction phase include force-majeure, such as earthquake, flood, fire, landslip, pestilence and diseases. Table 8.1 illustrates a number of typical construction risks.

A client organisation wishing to procure a project will need to identify the risks, not only during construction but also for the whole project life cycle. To ensure that risks are dealt with in the best way the client must determine the contract strategy best suited to a particular project's life cycle. Unless the client is entering into just one contract for the whole project, which is unlikely, consideration must be given to consistent allocation of risk both horizontally and vertically throughout the supply chain. That is, consideration must be given to the allocation of risks between designer and suppliers just as it is to interface problems between different suppliers. That is, communication and organisational issues need to be addressed both vertically and horizontally throughout the supply chain, unfortunately, this is taking place when there is the greatest amount of uncertainty about the project, see Figure 8.4.

Table 8.1   Typical construction risks.

Physical	Natural, ground conditions, adverse weather, physical obstructions
Construction	Availability of plant and resources, industrial relations, quality, workmanship, damage, construction period, delay, construction programme, construction techniques, milestones, failure to complete, type of construction contracts, cost of construction, commissioning, insurances, bonds, access and insolvency
Design	Incomplete design, availability of information, meeting specification and standards, changes in design during construction
Technology	New technology, provisions for change in existing technology, development costs and IPR and need for research and development

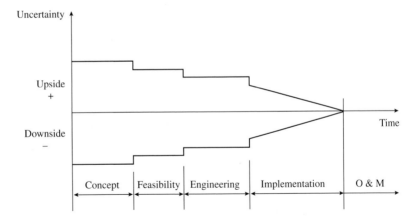

Figure 8.4 Project risk exposure.

Below a number of risk strategies are examined and their usefulness in a number of situations is discussed. Whatever approach is taken the implications for the whole cycle of the project must be considered and the goals of all parties aligned

Figure 8.5 illustrates the interrelationship between the basic management requirements of flexibility, incentive and risk sharing as provided by the different types of contracts described previously.

Conventional approach

This approach is commonly used in engineering projects. The parties' roles and responsibilities are based on the separation of design from

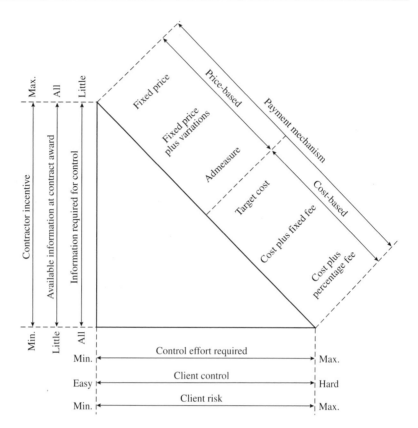

Figure 8.5 Factors influencing payment choice.

construction. The design is carried out by a consultant or in-house team, with limited (if any) contractor involvement, whereas the construction is the responsibility of the contractor with limited involvement of the client. The construction contract is usually supervised and administered by the engineering design consultant working on behalf of the client. As the parties responsibilities vary, their obligations, risk exposures and ability to carry risk also vary.

Contracts (normally admeasurements or lump sum) are usually awarded competitively, in most cases to the lowest bidder. Participants are usually pre-qualified and tenders are only invited from contractors after the design is complete. Co-ordination of design and construction is often the responsibility of a consulting engineering company. The engineer is obliged to provide information so as not to delay the contractor.

This organisational structure usually allocates the risk of changes in the price of items to the contractor while the risk of delay can be allocated to either the contractor or the client. The tendered prices used in this type of contract include a contingency for risk, which again means that the client is likely to pay more for the privilege of transferring the risks to the contractor than if he had accepted it himself.

Cost-based reimbursable approach

This form of contracting requires the client to take the majority of the risks as the contractor is paid on a cost plus fee basis but it also means that the client only has to pay for those risks that occur. The downside is that the client may pay for contractors' inefficiencies, which should be the contractors' risk. Cost reimbursable contracting can allow the contractor to have an input into the work at an early stage (leading to a high level of integration between design and construction) and this should help to reduce some of the project risks that may occur using a conventional approach. The more 'hands-on' approach adopted by clients usually leads to the contractor having a better understanding of the client's needs.

Management contracting approach

Management contracts are used by clients who want a third party to supervise and co-ordinate the design and construction of the project. These contracts require management contractors to place contracts for the packages of work and to oversee the project and ensure that the client receives what he initially specified. The client transfers all risks, except those associated with the operation of the project, to the management contractor. The management contractor can then transfer the risks that he holds through the contracts with the works contractors, as he wishes. The management contractor is usually reimbursed for all expenses he incurs, including those paid to sub-contractors. A fee (either fixed or a percentage of the total) is usually paid to cover overheads and profit.

Although the client carries most of the risk, this approach can avoid delays and claims by awarding smaller packages of work as the design is ready and it has the added advantage of flexibility and enhanced client involvement in the project.

This approach is often used in projects where there is a need for an early start and/or early finish, flexibility in relation to scope changes, special requirements relating to labour or construction methods and where the client has insufficient resources.

Fast-track approach

The fast-track method of construction requires the compression of the design and construction stages by the overlapping of many activities and although not a type of contract, fast-track is a method of constructing the works that requires a much greater degree of control over the construction process. Fast track projects are governed by contracts and it is necessary to choose a suitable type of contract to ensure that there is continuity in the work. This method of construction increases the risks in the project because the design of the work is not usually completed before the construction starts. If problems occur, they are less recoverable, from the programme point of view, than if using conventional methods of construction. This requires a large amount of co-ordination to ensure that the construction does not have to stop because the necessary designs are not completed. The use of fast tracking also means that the contractor must have a good relationship with his suppliers because he has very little advance warning of the exact quantities of goods that are required for a particular section of work. A management contract is often used for fast-track projects.

Turnkey/package deal approach

For this approach, the client gives detailed specifications of what he requires and awards a single contract for the entire facility. It is then the responsibility of the contractor to design, construct and commission the facility sometimes including operation and maintenance, and ensure that it conforms to the client's specifications. The contractor can sub-contract out the work, but it is the contractor who deals with the client. The client's involvement in a project of this type is minimal. These contracts can be termed, turnkey, design and build or package deal. A Build–Own–Operate–Transfer, BOOT, approach can be similar to this organisationally but in the case of BOOT projects finance has to be raised by the promoter which is repaid (in the form of tolls or tariffs) over a concession period, eventually the facility reverts to the ownership of the client organisation.

Turnkey arrangements are widely used by clients who know their requirements but have a shortage of in-house technical expertise and other resources necessary to carry out the work. They are also adopted when the design and method of construction are interrelated, where a patented process is required, where an early completion is desired or where sophisticated clients see that the most cost effective approach is to give all of the work to a specialist.

In preparing the contract, the performance, specification, standard required and sometimes the outline drawings for the preferred design are clearly stated. The contract should also clearly spell out the responsibilities and liabilities of both parties, especially those related to approval and acceptance of the design by the client before construction starts.

Usually in design and construct projects once the client approves the design he is responsible for any defects that may arise in it, while in turnkey projects the approval of the design by the client implies that it meets the specification. But this does not mean any technical approval or defect liability.

A package deal reduces the risks to the client compared with the conventional approach but increases the risks for the contractor. The contractor accepts all of the risks associated with the design and construction of the project while the client usually accepts the non-project risks and the risks associated with the operation of the facility. Although the client transfers the project risks to the contractor, once the specifications for the facility have been given it is very difficult for the client to make any changes or alterations.

This type of project is very inflexible for the client, despite the reduction in the amount of risks that have to be accepted. If the client wishes to make any changes or alterations when the specifications have been given, it will result in increased premiums and increase the chance that the project will not meet its objectives. For the contractor this type of project has increased the risks, but it does allow the contractor to use expertise and experience in planning and managing the work.

Normally, the contractor is paid on a fixed price basis there is no mechanism in this type of contract for price adjustments, so the price tendered by the contractor must include some allowance for changes in prices. The allowance included in the tendered price for price changes is the premium that the client pays for transferring the risk to the contractor.

This type of contract allocates the cost risk associated with the construction, and possibly the design and work to the contractor. There may be a clause in the contract that requires the contractor to pay the client in the event of a delay, but the inclusion of this clause is left to the discretion of the client. Otherwise, the risk of a delay in the project is retained by the client, along with all the other risks in the project. As the management of design and construction is integrated within one organisation this contract strategy approach has become attractive to clients, particularly when this is in conjunction with the ability to predict a firm price at an early stage of the project.

For this organisational system to be adopted effectively and in order to avoid potential disputes, it is necessary for clients to state their objectives clearly and to clarify their specifications and needs prior to contract commitment. Contract conditions included on offers should be considered carefully to avoid potential contradiction of the clients needs.

Framework agreements

A Framework agreement is a long-term commitment between the parties to enable clients to place contracts on pre-agreed terms, specifications, rates, prices and mark-up that are embedded in the framework to cover a certain type of work over a period of time or in a certain location or both. A framework agreement in itself gives no work to the contractor and may be non-exclusive. The contractor agrees to make staff, designers and construction resources available to undertake these contract packages as they are awarded and ensures their completion within agreed standards and timescales

As shown in Figure 8.6 a model framework agreement contains the following constituents – enabling elements; confidence in partner co-operation and ownership balance and structure. All three constituents are interrelated and exist in a multi-project environment or long-term relationship. The quantity and quality of the enabling elements affects the ownership balance and structure, and could raise or lower confidence in partner co-operation. Should the confidence in partner co-operation change, a party may deem it necessary to alter its strategic approach through its enabling element provision and this, could have a secondary effect regarding the ownership balance and structure. Likewise, adjustments to the ownership balance and structure has subsidiary effects on the other two constituents. In other words, as these elements change so does the balance of risk between the parties.

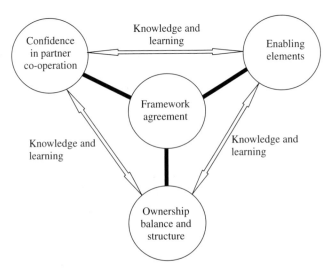

Figure 8.6 A model for framework agreements.

Framework agreement ownership is the balance of the privity of each partner relevant to its contribution or the division of the equity invested. If an agreement between two partners is equally split then this is an equal partnership. Any other balance will result in a majority–minority ownership. Both majority–minority and equal ownership can exist when multiple partners exist.

The ownership balance and structure can usually be related to the level of risk, commitment, provision of resources, the agreement, organisation structure, financial contribution and so on. Therefore, dependent on the negotiation procedure, the ownership balance and structure can govern what each partner must contribute towards the framework agreement and the allocation of risk between them. Knowledge and learning transfers underpin these relationships.

The enabling elements are goal compatibility, complementary resources, commitment, capability, financial traits, organisational traits, strategic traits and cultural traits. They identify the goals and objectives of the contractual arrangement and hence the optimum balance of risk. The enabling elements contain equitable/inequitable and tangible/intangible resources, goals, capability, knowledge and so on. If the goals of the partners are not collectively compatible, it would be more effective to address these goals individually rather than through a framework agreement as the agreement would be likely to fail.

Confidence in partner co-operation is also important. Should confidence be high, the framework agreement will run smoothly, attain all goals and objectives of the partners and the agreement. If confidence is low, goals may not be achieved, commitment to the agreement could reduce and peer pressure increase. This will be a reflection of the parties' ability to manage risk.

Trust has an influence on the confidence of partner co-operation and in turn the enabling elements and ownership balance and structure. Should trust increase, confidence in the framework agreement will increase and that party shall be more committed to the agreement. Increasing commitment means that the party will vary its organisational, strategic objectives, cultural adaptations and resource, compliment more to the partner's agreement, compatible goals and objectives.

A high level of trust will improve partner adaptability and the willingness to modify the ownership balance and structure for the benefit of the agreement and partners' needs. Without high confidence in partner co-operation, parties will be suspicious of any suggested alterations to the ownership balance and structure. This suspicious attitude could result in a breakdown in trust and confidence overall, thereby terminating the ownership balance and structure, together with the framework agreement.

Partnering

A good definition of partnering is provided by the Reading Construction Forum, in *Trusting the Team*, University of Reading, Reading (1995):

> Partnering is a managerial approach used by two or more organisations to achieve specific business objectives by maximising the effectiveness of each participant's resources. The approach is based on mutual objectives, an agreed method of problem resolution, and an active search for continuous measurable improvements.

This definition focuses on the key elements that feature prominently in partnering, irrespective of the form it takes namely mutual objectives, agreed method of problem resolution and continuous measurable improvements. Over the years the traditional construction relationship has lacked any degree of objective alignment, and provides for no improvement in work processes. Parties enter the project focused on achieving their objectives and maximising their profit margins, with little or no regard for the impacts on others. This mindset leads to conflict, litigation and sometimes a disastrous project. The characteristics of such a competitive environment includes objectives, which lack commonality and actually conflict, success coming at the expense of others (win or lose mentality) and have a short-term focus.

Partnering has been widely advocated for the industry in the United Kingdom to rectify the adversarial contractual relationships that have jeopardized the success of many projects, by improving collaboration and trust.

Features of partnering relationships have been seen in various industries for many years. The partnering style of relationships with contractors was a feature of some construction projects early in the Industrial Revolution. As applied today it originates in the philosophies of the Japanese-influenced automobile industry. The defence, aerospace and construction industries have followed. Its essence is the *alignment* of values and working practices by all members of the supply chain in order to meet the customer's real needs and objectives, though this has been pursued with different degrees of success and sustaining it is a questionable objective. Continuous improvement has been an important objective, with emphasis not only on cost but also on quality, lead time, customer service and health and safety at work. Incentivising the partnering companies by sharing cost savings has been a feature of continuous improvement and performance-based partnering in many industries, but in construction this has often been less significant than the primary objective of avoiding disputes.

The idea of alignment is significantly at odds with traditional practice in many industries. Procurement in most of the public sector has been based historically on accepting the lowest price bid. Much private construction also traditionally operated on this basis. It has led to conflicts about paying the actual costs of work, which revolve around risks and financial self-interest, between the various stakeholders – such as clients, design team, consultants, main contractors, sub-contractors and suppliers – throughout the construction process. As a consequence the final cost of the project usually exceeds the contract price and the result is confrontation.

The objective is to create a win–win culture so that projects are completed successfully and thus recover the confidence of clients in the industry. Partnering is a process to establish good relationships at all the interfaces between stakeholders and their commitment to the job and each other. Partnering should create trust, teamwork and co-operation to give early warning of potential problems and establish effective authority to agree decisions on them. It is critical throughout a project to remove traditional barriers and perceptions of unfairness between the parties involved. By changing to a win–win style the parties can reap benefits of cost saving, profit sharing, quality enhancement and time management. Unifying all the parties into one team for a project also reduces transaction costs.

Alliances

Alliances are another way of organisations working together to achieve mutual objectives. They have typically been used in the offshore industry where the previous convention had been to place separate and discrete contracts for engineering/procurement, deck/module fabrication, installation and hook up and commissioning with the contract reimbursement arrangements based on fixed price and all-inclusive rates (i.e. labour costs, overheads and profits, etc.) for the bulk of the work. The bulk of the risk was traditionally being placed at the contractor's door. The client, with a large team, ensured the quality, handled all the technical, contract interfaces/barriers and carried overall responsibility for success. The conventional approach required the scope of work to be well understood, fully defined and fixed, prior to contract placement. The contractors' objectives were thereby aligned to his scope and services which tended to be misaligned with the other contractors and the client. There had also been problems as design was being finalised during fabrication, leading to changes and cost increases in downstream contracts.

The alliance project execution strategy is based on the general principles listed below:

- ❑ aligned goals and objectives;
- ❑ commitment to aggressive cost reduction;
- ❑ working to remove process inefficiencies, duplication and traditional contract interfaces;
- ❑ formation of integrated teams;
- ❑ early involvement of contractors to optimise design and reduce costs;
- ❑ aligned and equitable contracts;
- ❑ shared profits and risks through individual contractor performance and overall alliance performance;
- ❑ promotion and maintenance of the highest standard of safety and quality.

The contractors are required to tender the expected out-turn total costs and accept earning profits through cost reductions and the need to achieve the client's facilities performance requirements based on pain/gain share related to balance of risk. This produced a culture enabling costs to be reduced (e.g. rather than use changes to gain profit) and total contractor buy in, coupled with the contractors ability to influence the topside development based on the client's requirements.

An alliance charter is a non-legal document of intent, developed to further amplify the alliance execution principles and to align and encourage joint working and promotion of relationships. This charter would normally be signed by the alliance members' very senior executive officers; thereby stating their commitment to the success of the project.

From a client perspective it is important to take on board the concept of achieving the lowest total life cycle cost which includes capital and operating costs. Industry can make a significant contribution to these goals and there is a potential for many and varied contributions. By the involvement of contractors early in the life cycle, with and/or through an alliance arrangement and solving problems in the early development phases, a marginal prospect could be turned into a viable project.

8.5 Risk allocation according to payment mechanism

According to this classification, there are two main categories: price-based and cost-based contracts. In the former the price and rates are submitted by the contractor in his tender. Lump sum (or fixed price) and admeasurements contracts lie under this category. The case is different for cost-based contracts where the contractor is reimbursed for the actual costs he incurs with a fee for overheads and profit. Cost reimbursable and target-cost contracts are in this category. Figures 8.3 and 8.4 should be referred to when reading this section.

Lump sum or fixed price

Some clients wish to transfer all of the construction risks to the contractor and be certain of his commitment. Usually, the responsibility for the package is vested in a single contractor, and bid evaluation is straightforward. The contractor agrees to carry out the work for the amount of money stated in the contract regardless of its actual cost as long as there is no change or breach of the contract from the client. This is quite common for schools, warehouses and similar works where the scope is relatively well defined and the work is straightforward.

Disputes are likely as there is very little scope for the incorporation of change and the client may well be required to pay the contractor more money in the long term. Even for the types of construction mentioned previously, the quantities and/or execution time may not be predictable. Unforeseen ground conditions, for instance, may affect the volume of excavation and type of foundations required. Extremes of weather may prolong the contract period. These risks that are outside the contractors control are usually the clients responsibility and there are often provisions in the contract to deal with them, although sometimes the price is absolutely fixed. Alternatively, some of these risks may be allocated to the contractor under the terms of the contract then an insurance premium is paid to him in the form of a risk contingency included in his bid. This consequently results in increasing bid prices.

From the contractors point of view fixed price contracts are a good opportunity to maximise profits. As the clients involvement in the project is minimal good planning, efficient use of resources and effective control can reduce costs and maximise profit. These contracts are normally let after competitive tender and so it is possible for the contractor to underestimate the costs involved. If this happens, he may not be able to cover the contract expense and, in extreme cases, he could become bankrupt. The quality of work and the programme may also suffer.

Another potential source of dispute relates to the degree of flexibility given to the contractor to determine the detailed specification. As long as he conforms to the information provided by the client then rejection by the client will put the client in a weak negotiating position and give the contractor the right to claim.

An implication of this approach is that contractors are responsible for risks over which they have little or no control, and of which their understanding and knowledge may be at best incomplete. However, when the client uses a fixed price format, project risks should be low, or they could be used when the client has little knowledge of a project's technological and construction requirements beyond the specification required. Those underwriting the project financially may not be comfortable with this approach, as it requires placing a large amount of trust in the contractor.

Admeasure

Admeasurement contracts require the use of a bill of quantities (BoQ) or schedule of rates. The work that the contractor is required to carry out is itemised, and it is necessary for the contractor to put a rate against each item of work. This method allows for the adjustment of price by the use of the itemised tendered rates. Uncertainty remains about the final price for the work because it does not necessarily follow that the lowest tender will be the one to give the lowest final price for the work.

Admeasure contracts are commonly used in building and civil engineering projects, especially in the public sector. They are usually used when risks are relatively low and quantifiable, the programme is almost fixed and the design (even when the quantities are approximate and subject to change) is almost complete and ready to be included in tender. In cases where design and construction need to be overlapped care should be exercised to ensure that there is sufficient information from which quantities can be obtained. Once the quantities have been inserted in the tender the contract conditions state to what extent these are subject to remeasurement. Amended detailed drawings may be issued while work is progressing but any considerable deviation in either quality or quantity of work from that forecast in the tender is a source of potential dispute on which contractors may base claims. Admeasurement contracts are more flexible than lump sum contracts because they allow for additional work to be priced using the contractors own pricing scheme. However, this only works to a limited extent, significant changes will lead to delays, disruption and claims.

Under this type of payment mechanism, the client's main concern when entering into a construction contract will be the risk that the duration or cost of the project will exceed their estimates. The contractor's principal risk is that undertaken when he makes an offer based on his tender estimate. He accepts the possibility of incurring greater costs than the income provided by his prices. The contractors can include contingencies in their tender anywhere that they choose, this being part of the skill of tendering.

However, this is limited because contingencies increase price and reduce competitiveness.

The successful application of the price based approach requires that the client and the contractor trust each other and that both parties have the same perception of the probabilities of occurrence and effect of the project risks. This is frequently not the case and confrontational events ranging from disputed claims to litigation result.

Cost reimbursable and target cost

As the name indicates the cost-based contract is one in which the contractor is reimbursed the actual cost he incurred carrying out the contract work plus a specified fee for overheads and profit. No total price is quoted at tender, competition is limited to the fee and technical capabilities, in most cases contracts are let after negotiation. Details of proposed management procedures and resources to be utilised must be given at tender. In the building industry, this tends to be known as 'fee contracting'.

Cost-based contracts have been used for process plant and some building and civil engineering contracts for several decades. The main use for cost reimbursable contracts is for projects where there is a need for an early start while the scope is not well defined. They are also used for works where the quantity of work is not well defined, demolition, site clearing, repair works or incomplete contracts where work was interrupted and for innovative works where research and development or novel design is required. Cost reimbursable contracts are appropriate in these cases because they are flexible; there is a high degree of client involvement through an active management role, which gives the client confidence that the contract will be properly executed. In order to maximise benefits and to ensure that the work is carried out efficiently and economically the client must maintain constant and detailed involvement in the project. The project team – in addition to technical and administrative supervision – must ensure that the contractor is utilising resources efficiently.

In order to protect the client's interests and to reduce the risks borne by him certain constraints must be imposed on the contractor through provisions in the conditions of contract. Unique clauses may be required for particular projects making contract administration more demanding, if these are not worded correctly they may cause unnecessary disputes.

Target cost contracts are based on setting a target, or probable cost for the work. The target cost set can be adjusted for any major changes in the work or for the cost of inflation. Payment for the work is similar to cost reimbursable contracts in that the contractor's actual costs are monitored and reimbursed. Any difference between the actual cost of the work and the target cost set for the work is shared, in a predetermined way, between the client and the contractor. There is also a separate fee, payable to the contractor, for overheads and profit. The target cost is usually arrived at through a two stage tendering process. An incentive is usually associated with the achievement of the target costs and so the main area of conflict when this contractual mechanism is utilised is the agreement of the adjustment of the target when the scope of work changes. Another significant risk is that the target is set at an unrealistic level. If it is too low the contractor must claim more and the client loses out.

These contracts have gone some way towards embracing the idea of the ideal contract, the most cost effective being the one that assigns each risk to the party that is best able to manage and minimize that risk, recognising the unique circumstances of the project. This approach takes the minimisation of the effect of risk as desirable, and allocates risk in accordance with this aim. A greater involvement is required from the client in return for potential construction cost savings. The risks in this type of contract are similar to those encountered in a cost reimbursable type of contract; however, there is added risk that in trying to finish at a cost lower than the target cost the contractor may use substandard materials. This risk would have to be borne by the client.

A target cost contract is often used for large projects where an early start/completion is required, before the design is complete or scope inadequately defined. It is also used for technically or organisationally complex projects where there is a need for the contractors' involvement in design and for projects with unquantifiable risks. In cost-based contracts, however, the payment of the contractor's costs means that the contractor is unable to inflate the direct cost of plant, labour or materials activities as these will be exposed when they submit their requests for payment. Thus, the principal location of contingencies in cost-based contracts is in anticipated plant utilisation and predicted rates of productivity. It has the significant disadvantage of high administration costs unless time is the target.

8.6 Contract award

The selection of external contractors is one of the crucial decisions made by the client if the project is to be a success. The criteria for selection may be price, time or expertise. The price criteria is often the key objective issue as the client seeks the most economic price for the development, whereas time and expertise criteria are often seen as being lesser objectives because of the need to expedite the construction programme and the need for good quality workmanship. The client's objectives in the tendering process were summarised earlier as follows:

❑ to obtain a fair price for the work, bearing in mind the general state of the construction market at the time; and

❑ to enter into an agreement with a contractor who possesses the necessary technical skill, resources and financial backing to give the client the best possible chance of the project being completed within the required time, cost and quality standards.

Before entering into the tendering process, the client must draw up a contract plan to determine the number of contracts into which the project will be divided. The basic consideration of this plan is the effect of the number of contracts on the client's management effort. More contracts will lead to more interfaces and greater management involvement, whereas fewer contracts reduce this involvement but may increase the client's risk exposure. There are certain principles that should be used when determining the number of contracts:

❑ the size of each contract should be manageable and controllable for the contractor;

❑ the contract size must be within the capacity of sufficient contractors to allow competitive tendering; and

❑ the time constraints of the work and the capacity restrictions allow for the separation of contracts rather than one single contract.

The tender process may take a number of forms; the main distinguishing feature is the level of competition. Open tendering involves a high-risk element for the client, as many of the tendering organisations will be unknown. With selective tendering in either one or two stages a limited number of organisations are invited to tender after some form of pre-selection or pre-qualification has taken place. In this case, award to the lowest conforming tender is not such a high-risk strategy. Negotiated tendering takes place when a client approaches a single organisation, based on reputation, but this can also be time consuming. The risk here is that at a later stage in the project the client may question whether or not VFM has been achieved in the absence of competition.

The tendering process has three key stages – pre-qualification, tender documentation and bid evaluation. A number of factors will influence the pricing policy of the tendering organisation, such as, competition, availability of resources and workload; however, these should not influence the criteria that the client uses for selection, but rather be taken into account as part of the client's evaluation.

A thorough pre-qualification should eliminate many of the risks that relate to the external organisation. These can be stated as:

- ❏ *Financial* the investigation involves an assessment of the financial statements, a check on the financial exposure of the company on domestic and overseas contracts and the history of recent financial disputes.
- ❏ *Technical* the assessments are concerned with the current commitment of labour and plant resources, the ability to handle the type of, quality and size of work at a specific time and performance on site for previous projects.
- ❏ *Managerial* the investigation involves identifying the managerial approach to risk, contract strategy, claims and variations.

Once the pre-qualification is complete, a list of tenderers can be compiled. These tenderers will then be issued with tender documentation, when this has been completed and submitted all of the tenders can be evaluated. Evaluation is primarily concerned with the justification of the lowest price bid that meets the client's overall requirements. It is essential that the client defines clearly and precisely the bid requirements to ensure that the submissions can be evaluated in terms of a common information base.

The detailed evaluation of technical, financial and contractual information can be carried out within the relevant departments of the client's organisation, with specific checks on technical expertise, price and contractual points being sufficient to identify the three lowest complying bids. Then the client should carry out a further evaluation and, if necessary, evaluate any alternative bids.

The construction programme submitted by the contractor should be taken into account as part of the technical evaluation as it indicates the contractors overall approach to the work, although many standard forms do not require a programme to be submitted at tender. The method statements and the proposals for plant and labour levels can be submitted as non-contractual information and used to justify the bid. Often this information is used to reject infeasible bids rather than to select the optimum bidder. The client must ensure that the technical information requested from the contractor is relevant to the evaluation procedure.

Although the bid price may be considered as the dominant factor, it is essential that all of the financial implications be examined. The evaluation process begins with an arithmetic check on a rate-by-rate basis identifying discrepancies between each bid and relative to the clients estimated rates. In some cases, one tenderer may load a particular rate based on his own calculations of the quantities required. The client may consider that some rates have been deliberately front loaded and request the tenderer to reallocate those rates without prejudice to the overall bid price. The client may also compare the estimated cash flow projections with those of the tender bids, occasionally using a discounted cash flow analysis as a further evaluation parameter.

The contractual evaluation may be carried out as a separate assessment or as a part of the technical and financial evaluation. Compliance with the contract documents are considered to be paramount. Any qualifications included in the contractors bid that had been accepted in the initial evaluation stage would be re-examined and clarified, if necessary, the bid may have to be rejected. The contractual evaluation is summarised in a report identifying those areas of risk and possible contractual problems associated with each of the bids.

8.7 Summary

When developing a contract strategy it is important for the client to communicate to the contractor his objectives, this may be done through a meeting (sometimes referred to as an alignment meeting) or it may be done through the contract documents but the message must be clear and unambiguous. The planning of the realisation of the project must be thorough and consistent through all aspects of work and functions involved to ensure that the objectives can be met successfully. Any targets set for contractors in terms of time or cost ceilings must be realistic and deliverable so that both parties can clearly see that they have been met.

In terms of entering into a contractual relationship, the client must ensure that the most appropriate risk sharing strategy is chosen and reflected by the organisational structure and payment mechanism adopted. The contract should then go on to define the obligations and rights of every party. In determining the risk allocation and so contract strategy, it is important to apply risk analysis and management techniques to ensure that the worst-case scenario has been anticipated and provision has been made to deal with risk events as and when they occur. Choose the terms of contract logically, depending upon the nature of the work, its certainty, its urgency, the motivation of all parties and other factors such as the relationship between an investment in new plant and systems already in use.

Time, money and resources are expended by many clients at the appraisal stage of a project because no attention is paid to the contract strategy available to cover the project risks as well as the project objectives and hence value.

Chapter 9

Managing Financial Risks in Major Construction and PFI Projects

Many projects are considered high risk solely because of the amount of uncertainty involved in the method of financing. It is therefore necessary to increase the chance of a project's success by identifying the risk associated with the finance and taking the necessary risk management actions. The risks associated with financing major projects, whether by public or private organisations can be numerous. All client organisations need to consider the risks typically associated with major projects, those being construction and operation risks and the risks associated with how the project is to be financed. The purpose of this chapter is to identify the different types of finance, appraisal and validity of finance and the risks borne by clients, promoters and lenders.

9.1 Project financing

Project finance is the term used to describe the financing of a particular legal entity, whose cash flows and revenues will be accepted by the lender as a source of funds from which the loan will be repaid. Thus, the project's assets, contracts, economics and cash flows are segregated from its promoter's such that it is a strictly limited recourse, in that lenders assume some of the risk of the commercial success or failure of the project.

Conventional project finance provides no recourse and if project revenues are insufficient to cover debt service, lenders have no claim against the owner beyond the assets of the project, the project, in effect being self-funding and self-liquidating in terms of financing.

Unlike traditional public sector, projects whose capital costs are largely financed by tax revenues or loans raised by government, concession projects are typically financed by a combination of debt and equity capital, the ratio between these two types of capital varies between project and country.

Every project requires finance whether it is a public, public/private or private project. The sources of finance for clients in the public sector are usually through taxation or grants. In recessionary times, only urgently required projects gain funding as income from tax revenues becomes limited. Grants, often in the form of interest-free loans can make up the project finance required.

Many projects realised in developing countries have been funded through multilateral, bilateral or tied funding. In most cases, the lack of sophisticated money markets has resulted in funding being restricted to a limited number of sources. The risk of repayments in exotic currencies where there is no active exchange market often deters lenders from funding projects in developing countries.

Over the last decade, public and private partnerships have provided finance for a number of projects. Sources of finance have been from pension funds, insurance companies, commercial banks, niche banks, large corporations, stock markets, aid agencies, developers and vendors.

Privately funded projects fall into two categories; those being concession type projects awarded by government and private developments. Sources of finance for these projects are often similar to those given below. The role and involvement of project and private finance in public services continues to extend past those boundaries traditionally associated with off-balance non-recourse financing. Many recent developments from the mid-1990s have revolved around public–private partnerships centring upon balance sheet treatment and limited recourse structures.

The satisfactory completion of several private finance initiative/public–private partnerships (PFI/PPP) projects has also led to the improved predictability of project cash flows, resulting in a tightening of debt service ratio requirements, lowering of risk premium placed upon debt instruments and lengthening of the term of loans available. This has been combined with specific financing strategies enabling project finance to be replaced with structures resembling those of a corporate finance nature, especially as projects reach expected operational targets. Takeout financing, where long-term bonds are used to replace original debt financing often referred to as refinancing or restructuring is a common strategy for the transfer from project to corporate financing. However, several markets that operate PFI/PPP forms of projects outside of the United Kingdom may not sustain the same degree of deal flows and liquidity in secondary markets.

9.2 Types of finance

The financial plan will almost certainly have a greater impact on the terms of the contract than the physical design or construction costs. In many projects, the lender may be a commercial bank, a pension fund, an insurance company, an export credit agency or a development bank. The structure of the loan may be in the form of debt, loan and debentures.

When structuring the terms and conditions of the loan for PFI/PPP projects that are contract-led (projects with a defined revenue structure dependent upon the completion of specific conditions, such as output specifications and performance standards) lenders may seek to limit the amount of deductions to the predicted revenue stream. The negotiated levels of deductions possibly and probably made may form the upper limits with regard to the total quantity of debt lenders are prepared to commit to the project.

Alternatively, market-led projects have to identify the degree of variance associated with future revenue streams and allocate appropriate quantities of equity (risk capital) before lenders may commit debt. Debt can be sculpted to address benefits associated with inflation as per tolled motorway systems; however, the acceptance by the lenders are often limited to those markets with experience of such deals or have access to lenders prepared to lend on a long-term basis.

The conditions of loan finance depend on the criteria of the lender and the promoter; the type and characteristics of projects considered and their location. The main features that need to be agreed are the type of loan and the repayment method. A mortgage may take the form where repayments are constant while capital and interest vary. Initially this means that the proportion of capital repaid is small but progressively increases throughout the project. Such structures may be difficult to create especially where capital expenditure is incurred throughout the concession, for example, the replacement of rolling stock, refurbishment or replacement of power turbines. In such instances, a rolling credit agreement may be more suited to allow the continued adjustment of the capital and interest calculation.

With equal instalments of principal, the amount of principal repaid is constant with each payment with the total amount paid decreasing over time. Under the concept of maturity, the principal is repaid at the end of the loan period in one sum often referred to as sunset or bullet payments. This form of loan is most suitable for projects, which generate a large capital sum on completion. Other variations on repayment structures may include a moratorium on capital repayments or interest payments for a period at the start of the loan.

In conventional loan finance the lender seeks to limit risk by insisting on the borrower providing security for the land, this is usually in the form of a

Redeemable preference shares are shares giving the investor the right to a fixed return and to obtain repayment of the investment at the end of a stated period.

Convertible preference shares are shares giving the investor the right to a fixed return and to convert to ordinary share capital at a future date.

9.3 Appraisal and validity of financing projects

The financial viability of a project must be clearly demonstrable to potential investors and lending organisations. In assessing the attractiveness of a financial package, project sponsors should examine the risks associated with the following elements:

- interest rate, debt/equity ratio (percentage being financed), cost of capital, exchange rates, acceptable margins;
- repayment period, currency of payment, associated charges (legal, management and syndication fees), securities (guarantees from lenders) and documentation (required for application, activate and drawdown of loans); lenders require security, sponsor support, cash deficiency guarantees, sinking funds, covenants, project finance documentation and inter-creditor agreement;
- default triggers especially those linked to debt service coverage ratios, loan life coverage ratios, interest cover ratios, working capital availability, standby loan facilities, contingency funds, debt and maintenance service reserve accounts;
- credit rating organisations can be brought into assess the risks in the project, which may result in financial default; loan stock or financial instruments issued based upon a projected cash flow may be rated based on a risk assessment methodology. Several tertiary parties offer such services such as Standard & Poor's, Moody's and Fitch.

The three basic financial criteria against which success needs to be measured in projects are:

- finance must be cost effective, as far as possible;
- the skilled use of finance to immunize the negative implications of risk associated with interest and inflationary fluctuations; and
- finance should be required over a term that provides acceptable refinancing horizons.

The project must have clear and well-defined revenues that will be sufficient to service capital and interest payments on the project debt over the term of the loans and to provide a return on equity that is commensurate with development and long-term project risk taken by equity investors. The credit of the borrower and the type of project needs to be considered by the lender in determining the type and value of loan required. This can be evaluated by a thorough examination of the borrower's financial status, track record and familiarity with specific types of projects.

For example, investment banks will normally fund infrastructure projects for a period of up to 25 years and industrial and process plants up to 14–17 years because of the cycle time before major maintenance and refurbishment is required. Institutional investors such as insurance companies and pension funds consider projects with fixed rates of return upto 20–40 years to match the cash flow characteristics of their liabilities. Lenders often refer to a robust finance package as one, which allows repayment of loans under a worst-case scenario over the loan period.

When selecting the sources and forms of capital required, the strength of the security package, the nature of the country risks and limits, the sophistication of local capital markets should be considered. One potential risk is that repayments are in inconvertible currencies, which cannot be exchanged, either because this is forbidden by foreign exchange regulations or because there are no buyers who wish to acquire the currencies.

One of the most important elements to be satisfied in a project is how to provide security to non-recourse or limited recourse lenders. If a promoter defaults under a project strategy utilising a non-recourse finance package, the lender may be left with a partly completed facility, which has no market value. To protect lenders therefore, various security devices are often included to protect the lender, these may include:

❏ revenues are collected in one or more escrow accounts maintained by an escrow agent independent of the promoter company;
❏ the benefits of various contracts entered into by the promoter, such as, construction contract, performance bonds, supplier warranties and insurance proceeds will normally be assigned to a trustee for the benefit of the lender;
❏ lenders may insist upon the right to take over the project (step in clauses) in case of financial or technical default prior to bankruptcy and bring in new contractors, suppliers or operators to complete the project;
❏ lenders and export credit agencies may insist on measures of government support, such as, standby subordinated loan facilities which are functionally almost equivalent to sovereign guarantees;
❏ lenders may insist upon monoline insurance of less senior instruments adopted for project finance;

❏ lenders may require risk coverage of specific financial risk or t
involvement of multilateral or syndicated loan facilities to reduce
political default.

The contract between the borrower and lender can only be determined
when the lender has sufficient information to assess the viability of a
project. In most projects, the lender will look to the project itself as
a source of repayment rather than the assets of the project. The key
parameters to be considered by lenders include:

❏ *total size of the project*: the size of the project determining the amount
of money required and the effort needed to raise the capital, internal
rate of return on the project and equity;
❏ *break even dates*: critical dates when equity investors see a return on
their investments;
❏ *milestones*: significant dates related to the financing of the project;
❏ *loan summary*: the true cost of each loan, the amount drawn and the
year in which drawdowns reach their maximum;
❏ host country guarantees to cover repayment of debt (where
applicable).

A properly structured financial loan package should achieve the following
basic objectives:

❏ maximise long-term debt;
❏ maximise fixed rate financing;
❏ minimise refinancing risk.

Financial closure is also a critical element to securing the financing, as it
is the last risk-mitigating step before the finance is allocated to the project
by shareholders and investors. Financial closure assesses the project's
technical, financial and legal probity and robustness.

It is important to realise that the financial plan may have a greater
impact on the terms of a project than the physical design or construction costs. Many technically sound projects have failed commercially due
to their financing structure especially during the operation stage when
revenues from users have not met those predicted.

If a country is in deep recession, then it is highly likely that only urgently
needed projects will be financed from the public purse. Public client
organisations are accountable to the public to ensure that projects are
brought in on time to budget and must ensure that they have sufficient
funds available to meet the cost of the project and to ensure that contingencies are available, should they be needed. The risk of over spending on
a project can result in other urgently needed infrastructure being delayed.

ntractors are often required to price projects very keenly as competi-
ı is great. Clients often suffer from claims-conscious contractors to try
neet margins of profit.

ı a strong booming economy, there are often increases in taxation revenues, which permit more projects to be undertaken and funded through
the public purse. During these periods, clients run the risk of poor quality
and paying over the odds for projects.

In developing countries, project finance may often be realised from
multilateral or bilateral lenders or may be in the form of tied loans.
Many developing countries run the risk of not being able to meet repayments due to economic conditions or overspend. In the case of tied
loans, clients run the risk of being tied to products from the donor's
country, which are not necessarily the best ones to meet the projects
objectives.

9.4 Typical financial risks

Financial risks are common to most projects. In some cases, the financial
risks are dependent on the occurrence of other risks, such as, delay in
construction or reduced revenue generation.

Typical financial risks include:

Interest: type of rate, fixed, floating, capped, floors or collars, changes in
interest rate, existing rates;

Payback: loan period, fixed payments, cash flow milestones, discount
rates, rate of return, scheduling of payments;

Loan: type and source of loan, availability of loan, cost of servicing loan,
default by lender, standby loan facility, debt/equity ratio, holding period,
existing debt, covenants;

Bond: Marriage of the bond drawdown to capital expenditure, issuing
time frames, uptake of the bond, credit rating of the bond, sufficiency of
the bond and standby facilities;

Equity: institutional support, take up (guaranteed or market dependent),
type of equity offered;

Dividends: timing and amount of dividend payments;

Currencies: currencies of loan, bands, ratio of local/base currencies,
depreciation and devaluation of currencies;

Market: changes in demand for facility or product, escalation of costs of
raw materials and consumables, recession, economic downturn, quality

of product, social acceptability of user pay policy, marketing of products and consumer resistance to tolls;

Reservoir: changes in input source;

Currency: convertibility of revenue currencies, fluctuation in exchange rates, devaluation;

Refinancing: liquidity of and sources for the secondary market, availability of instruments, improved efficiency and transactional charges.

Both borrowers and lenders need to adopt a risk management programme. Risk management should not be approached in an ad hoc manner but structured. The five major steps of such a process are:

(1)   identify the financial objectives of the project;
(2)   identify the source of the risk exposure;
(3)   quantify the exposure;
(4)   assess the impact of the exposure on business and financial strategy;
(5)   respond to the exposure, adapt the financial strategy and reiterate as required.

The first stage, is to develop a clear understanding of the project. Borrowers and lenders need to determine their objects regarding the financing of a project. Many borrowers seek long-term loans with repayments made from revenues. The risk of not meeting repayments is often reduced when the borrower has sufficient earnings at the start of operation to service the debt. Many projects, however, suffer commissioning delays, which increase the borrowers loans and repayments. In many cases, borrowers will seek grace periods from lenders to cover the risk such as delays.

Lenders seek positive cash flows and must ensure that their objectives are met by providing the best loan package. If a short-term loan is the lenders objective then the major risk will occur at the start of operation and should the project not generate sufficient revenues, the lender may need to consider debt for equity swaps, as in the case in the Channel Tunnel project.

Once the project objectives are defined, the overall costs, including construction and operation costs are determined, a cumulative cash flow model is prepared. The model can be used to quickly estimate the NPV, IRR, cash expenditure and payback period of a project. This model is initially prepared without considering potential risks. It is essential that the estimates and programmes prepared are reflective of cost and time over the projects life cycle. The risk of inaccurate estimates based on fixed budgets often lead to optimistic cash flows, which do not truly illustrate the effects of risk occurring during a project.

In many cases, the cost of finance is not included in the cash flow at this stage. Many organisations prefer to use the return on investment (ROI) as the measure of profitability. The authors, however, consider that the cost of finance along with all other projected costs and revenues should be incorporated in the cash flow as this provides a more accurate illustration of the projects finances. Working capital should also be considered in the cash flow as certain risks may occur and result in further borrowing over and above that estimated.

9.5 Promoter

In most projects involving the construction of public utilities such as highways, bridges, tunnels, power stations, factories and leisure centres, the government or one of its departments will authorise the project, which will often require special legislation and specific government approvals. There are many and varied roles to be considered including equity taker, loan or loan guarantor, provider of grants, new and existing facilities and of raw materials or feedstock, guarantee offtake quantity and price, completion guarantees, fiscal relief's such as tax, duties, social contributions and political guarantees like non-cancellation, non-competition rights to toll existing facilities.

Promoter–lender: debt financing contract

The contract between the promoter and lender can only be determined when the lender has sufficient information to assess the viability of a project. In projects, the lender may look to the project itself as a source of repayment rather than the assets of the project. The lender would consider the total size of the project – the size of the project determining the amount of money required and the effort needed to raise the capital, internal rate of return on the project and equity, the break even dates – critical dates when equity investors see a return on their investments, the milestones – significant dates related to the financing of the project and the loan summary – the true cost of each loan, the amount drawn and the year in which drawdown reach their maximum.

Promoter–investor: equity financing contract

In developed countries, a significant amount of equity can be raised for projects from investors in the domestic market, by means of either floating the project company on the stock market or through raising funds through private placement. In developing countries however, it may, be difficult to raise equity in the capital market, which will often result in debt instruments being utilised. Often equity participants will include the promoter, constructor, operator, major vendors and sub-contractors, the host country and passive investors looking for sound investment opportunities. However, the amount of equity participation should not result in the promoter losing management control of the project.

9.6 Financial risk in concession contracts

In the context of concession projects there are two types of risk, those being elemental risks and global risks, defined as:

(1) Elemental risks are those risks which may be controlled within the project elements of a concession project.
(2) Global risks are those risks outside the project elements, which may not be controllable within the project elements of a concession project.

In concession projects, for example, a negligible risk may be a risk associated with the technology utilised to meet the facility and a catastrophic risk may be the risk of expropriation by the principal before revenues are generated.

Risk management is not a discrete activity but a fundamental of project management techniques and the responsibility of the complete project team. In concession projects the project team representing the promoter need to determine the risks associated with each contract prior to appraisal.

The risk management criteria of a promoter organisation involved in gas-fired power concession project, for example, could be summarised as: the project involves a demonstrated technology and application, the principal or host country is an acceptable credit risk, verifiable equity returns are commensurate with risk, the project is environmentally sound, the project is compatible with the off taker(s) needs, partners with complementary strengths are available to participate and an acceptable market lender for the type of project exists.

The risks associated with concession projects need to be identified, appraised and allocated through a risk management structure, which addresses all those risks over the lifecycle of a project.

Although risk management is undertaken at the earliest stages of the project, when looking along the project lifecycle, there are two phases, when risks associated with financing concession projects occur, those being the construction phase and the operation phase.

These two distinct phases are considered as:

(1)   the pre-completion phase relative to construction risks;
(2)   the post-completion phase relative to operational risks, with the first few years of operation being the major operation risk.

Financial risk, political risk and technical risk must be considered as major elements of concession projects as are pre-completion and post-completion risks. Political risks may adversely affect the facility during either of these phases. Specific risks may be broken down into two main categories those being: global risks which include political and legal risks, and elemental risks which include construction and operational risks.

Promoters of concession projects are exposed to risks throughout the life of the project, which may be typically summarised as:

❑   development risks associated with competition and the concession contract;
❑   realisation risks associated with the construction contract and force-majeure;
❑   operation risks associated with revenue risk and cost and supply risks;
❑   country commercial risks:
  ■   convertibility and foreign exchange risks;
  ■   short-term development risks, medium-term construction risks and long-term operating risks;
❑   political risks:
  ■   political stability, profit making risks and default risks.

Risks associated with market prices, financing, technology, revenue collection and political issues are major factors in concession projects. Other risks encountered in concession projects may include physical risks, such as, damage to work in progress, damage to plant and equipment and injury to third persons; theoretical risks, such as, contractual obligations, delays, force-majeure, revenue loss and financial guarantees.

Since loans need to be repaid from the revenues generated by the project during the concession period the question of sovereign and exchange rate, risks will need to be considered beyond the completion of the construction phase.

Typical risks for a concession project include:

- *completion risk*: the risk that the project will be completed on time and to budget;
- *performance and operating risk*: the risk that the project will not perform as expected;
- *cash flow risk*: the risk of interruptions or changes to the project cash flow;
- *inflation and foreign exchange risk*: the risk that inflation and foreign exchange rates affect the project costs and revenues;
- *insurable risks*: risks associated with equipment, plant, commercially insurable risks;
- *uninsurable risks*: force-majeure;
- *political risk*: risks associated with sovereign risk and breach by the principal of specific undertakings provided in the concession agreement;
- *commercial risk*: risks associated with demand and market forces.

Sovereign risk, often associated with credit worthiness is a major factor in overseas concession projects. Sovereign risk does not apply in domestic lending where funds are sought from domestic banks that are able to better assess and absorb risks. Currency risk and revenue collection risk are also considered as major risks in overseas concession projects, since fluctuations in exchange rates and the ability to collect revenues may affect the commercial viability of the project.

Demand risks associated with infrastructure projects are much greater than those facilities producing a product off take since an infrastructure project is static and cannot normally find another market, whereas a product may be sold to a number of off takers through the life of the concession. However, facilities producing an off take bear the risk of product obsolescence and competition usually leads to market risks dominating, especially when operation and maintenance costs are high and concession periods short.

Construction risks, political risks, technical risks, financial risks and logistical risks are associated with construction contracts in developing countries. The logistical risks are:

- embargo;
- availability of spares, supplies, fuel and unskilled and skilled labour resources;
- loss or damage in transportation of materials and equipment;
- access and communications.

The type of facility, its location, method of procurement and method of revenue generation will determine the risks specific to a concession project. Each project will therefore be subjected to different types of risk.

The most serious effects of risks in Concession projects are:

- failure to meet cost estimates;
- failure to achieve completion dates;
- failure to achieve faulty and operational requirements determined by the concession agreement;
- failure to achieve estimated revenues;
- failure to meet repayments;
- failure by one party to meet his obligations as determined by the concession agreement.

Although risks may seem to imply a loss, they may also have beneficial effects such as:

- costs may be lower than estimated;
- completion may be sooner than anticipated;
- quality may be achieved at a lower cost while still meeting operational requirements;
- increase in demand increases the estimated revenue;
- increased revenue benefits repayments;
- obligations are met by each party to the agreement.

9.7 Global and elemental risks in concession contracts

In this section, concession project risks are identified and classified into two categories detailed further.

Global risks and elemental risks

The four major global risks are political, legal, commercial and environmental risks. The effect of global risk may be considered as shown in Table 9.2.

Many of the global risks identified may be addressed by provisions in the concession agreement. For example, minimum demand guarantees may be sufficient to cover many of the commercial risks; agreements on tariff or toll increases during the concession period may cover the major political risk. The cost effects of changes in environmental legislation may be covered by the provision to extend the concession period to cover such costs and legal requirements for the payments of duties may be relaxed.

Table 9.2   The four major global risks and their effects.

Political risks	
Concession	Delay in granting concession, concession period, price setting by principal, public inquiries, enabling bill, commitment to concession contracts, exclusively of concession, competition from existing facilities
Legal risks	
Host country	Existing legal framework, changes in laws during concession period, conflicting community, national or regional laws, changes in regulations regarding importation and exportation, changes in company law, changes in standards and specifications, commercial law, liabilities and ownership, royal decrees
Agreement	Type of concession agreement, changes in obligations under legal framework, changes in provisions of agreement, statutory enactments, resolution of disputes
Commercial risks	
Market	Changes in demand for facility or product, escalation of costs of raw materials and consumables, recession, economic downturn, quality of product, social acceptability of user pay policy, marketing of product and consumer resistance to tolls
Reservoir	Changes in input source
Currency	Convertibility of revenue currencies, fluctuation in exchange rates, devaluation
Environmental risks	
Sensitivity	Location of project, existing environmental constraints, impending environmental changes
Impact	Effect of pressure groups, external factors affecting operation, effect of environmental impact, changes in environmental consent
Ecological	Changes in ecology during concession period

The four major elemental packages associated with concession projects are construction, finance, operation and maintenance, and revenue generation. Each package will contain discrete components that make up the total package with assumed elemental costs and risks. These may include risks shown in Table 9.3.

The elemental risks classified are typical risks associated with specific project conditions found in concession projects. Many elemental risks are often risks that may be controlled by the promoter either directly or by contractual agreements with constructors, operators or lenders.

There are a number of occurrences which impact projects that are calculable risks, those related to construction projects being dynamic or

Table 9.3   The four major elemental packages associated with concession projects.

Construction risks

Physical	Natural, pestilence and disease, ground conditions, adverse weather conditions, physical obstructions
Construction	Availability of plant and resources, industrial relations, quality, workmanship, damage, construction period, delay, construction programme, construction techniques, milestones, failure to complete, type of construction contracts, cost of construction, insurances, bonds, access, insolvency
Design	Incomplete design, design life, availability of information, meeting specification and standards, changes in design during construction, design life, competition of design
Technology	New technology, provision for change in existing technology, development costs

Operational risks

Operation	Operating conditions, raw materials supply, power, distribution of off take, plant performance, operating plant, interruption to operation due to damage or neglect, consumables, operating methods, resources to operate new and existing facilities, type of O&M contract, reduced output, guarantees, underestimation of operating costs, licences
Maintenance	Availability of spares, resources, sufficient times for major maintenance, compatibility with associated facilities, warranties
Training	Cost and levels of training, translations, manuals calibre and availability of personnel, training of Principal's personnel after transfer

Financial risks

Interest	Type of rate, fixed, floating or capped, changes in interest rate, existing rates
Payback	Loan period, fixed payments, cash flow milestones, discount rates, rate of return, scheduling of payments
Loan	Type and source of loan, availability of loan, cost of servicing loan, default by lender, standby loan facility, debt/equity ratio, holding period, existing debt, covenants
Equity	Institutional support, take-up of shares, type of equity offered
Dividends	Time and amounts of dividend payments
Currencies	Currencies of loan, ratio of local/base currencies

(Continued)

Table 9.3   (Continued)

Revenue risks	
Demand	Accuracy of demand and growth data, ability to meet increase in demand, demand over concession period, demand associated with existing facilities
Toll	Market-led or contract-led revenue, shadow tolls, toll level, currencies of revenue, tariff variation formula, regulated tolls, take and/or pay payments
Developments	Changes in revenue streams from developments during concession period

speculative risks such as political, social, economic, environmental and static risks such as acts of God and man, damage to property, injury or production loss or gain. These risks can be dealt with through a formal or informal risk management process consisting of identification, estimation, analysis of alternatives and implementation.

The need for risk analysis seems particularly apparent when projects involve:

❏ large capital outlays;
❏ unbalanced cash flows requiring a large proportion of the total investment before any returns are obtained;
❏ significant new technology;
❏ unusual legal, insurance or contractual arrangements;
❏ important political, economic or financial parameters;
❏ sensitive environmental or safety issues;
❏ stringent regulatory or licensing requirements.

All, or a combination of a number of the above parameters are fundamental to concession project strategies and each risk identified in the project must have a uniform basis of assessment which will inevitably involve cost and time.

Since the revenue must be sufficient to service the debt, the total cost of the project must be reasonably predictable before analysis and to ensure lenders and investors are prepared to accept the risks identified in the project.

Project analysis of a concession project should consist of a number of analyses with the financial analysis taking the central role. This project analysis is summarised as:

Financial market analysis: this analysis considers data regarding the availability, cost and conditions of financing a project;

Cost analysis: this estimates the development, construction and operating costs and establishes a minimum cost of the project;

Market analysis: this forecasts demand and establishes a maximum price and evaluates the commercial viability of the project;

Financial analysis: this compares the cost, market and financial market analysis and establishes the relationship between costs and revenues.

Once the project risks are identified then analysis of the project should include:

- an assessment of the validity of the underlying assumptions made regarding the risks;
- test the sensitivity of the projected cash flows to the technical and economic assumptions, in particular the assumptions made about market risks such as future sales, prices and competition.

There is nothing new about risk analysis and most risks can be diluted by distributing them over many contracts and passing them onto the client. In concession projects however, the risks borne by the promoter and those allocated to other parties will influence the success of the project since the revenue generated over the concession period may suffer if risks are not sufficiently analysed and allocated.

A typical response to risks in concession projects may be summarised as:

Completion risk: cover by a fixed price, firm date, turnkey construction contract with stipulated liquidated damages;

Performance and operating risk: cover by warranties from the constructor and equipment suppliers and performance guarantees in the operation and maintenance contract;

Cash flow risk: cover by utilising escrow arrangements to cover forward debt service, guard against possible interruptions and take out commercial insurance;

Inflation and foreign exchange risk: cover by government guarantees regarding tariff adjustment formula, minimum revenue agreements and guarantees on convertibility at certain agreed exchange rates;

Insurable risks: cover by form of insurance such as policy to cover cash flow shortfalls mainly during the pre-completion phase of a project;

Uninsurable risks: cover by insisting host government provide some form of coverage for uninsurable risks such as force-majeure;

Political risks: cover by political risk insurance from export credit agencies or multilateral investment agencies;

Commercial risk: cover by insurance policies such as export credit guarantee department (ECGD).

Initially particular risks may be avoided by the promoter by determining the most suitable technical parameters to meet the requirements of the concession, the facility and its location. Second, a detailed analysis of the ability to replace, at no additional cost, the operation or finance element of the project. Third, risks would be transferred through the concession agreement to the principal and the contract with each organisation involved. Finally, the risks retained by the promoter would be those risks, which could not be allocated to other parties, such as, commercial risk or could not be insured against due to the premiums required.

A central feature in concession project financing is the allocation of risks and rewards to the parties willing to bear them. Having reached a stage where all major risks have been identified and analysed a specific structure may be formulated to achieve the financial objectives acceptable to potential lenders and the principal. However, problems occur when allocating risks between the principal and the promoter since transferring all risks will invariably result in increased prices thereby jeopardising the interests of the users and since there are no guidelines for risk allocation decisions are often subjective.

The concession represents a mechanism for the allocation of risks in concession projects and risk should be shared between the promoter and the principal through the concession agreement with constructor and operator risks covered by performance guarantees, completion guarantees, warranties and operating guarantees. Political risk should also be transferred through the concession agreement to the principal.

Many of the risks associated with construction and operation may be protected by performance guarantees, completion guarantees, warranties from suppliers, operating guarantees and regular inspection by the principal.

The involvement of off takers, vendors and contractors in a promoter consortium should allow the allocation of risk to those parties best able to manage it. For example, guarantees in off take contracts can be used to transfer risk due to changes in market conditions from the project users; take-or-pay contracts guarantee the project a future stream of revenues. Lump sum or turnkey contracts can be used to transfer completion and cost overrun risk to contractors. Performance guarantees and incentives in purchase agreements and operation and maintenance contracts can be used to transfer operating risk to suppliers and operators. Involvement of the government can also be used to manage political risk.

9.8 Summary

The risks fundamental to major and/or concession projects are far greater than those considered under a traditional contract strategy. The effects of risks on the project and the allocation of such risks to each organisation via the concession agreement and secondary contracts, over the different phases of the concession period, need to be identified and responded to, at an early stage of bid preparation. Elemental and global risks identified in concession projects need to be appraised in a logical manner and allocated through contracts and agreements to relevant parties. Risk analysis techniques need to be compatible with the nature of the project.

Chapter 10

Risk Management at Corporate, Strategic Business and Project Levels

10.1 Introduction

The increasing pace of change, customer demands and market globalisation have put risk management high on the agenda for forward thinking companies. It is necessary today to have a comprehensive risk management strategy. In addition, the Cadbury Committee's report on corporate governance (1992) states that having a process in place to identify major business risks is one of the key elements of an effective risk management system. This has been since extended in the guide for directors on the combined code, published by the Institute of Chartered Accountants (1999). This report is referred to as the 'Turnbull Report' (1999) for the purposes of this chapter.

The Turnbull Report is a timely reminder of the need for effective risk management and also creates an opportunity to review what an organisation has in place and to make the appropriate changes. Risk management can be considered as crucial to the sustainability of a business in its environment. In the past, large corporate failures have occurred where risk management was never even considered.

Reichmann (1999) states 'One of the most important lessons I have ever learnt, and I didn't learn it early enough, is that risk management is probably the most important part of business leadership'.

However, organisations do need to be pragmatic. Taking risk is often needed in order to gain reward. This is clearly stated in the Turnbull Report, which states that, 'risk management is about mitigating, not eliminating risk', and the board of directors of an organisation have the overall responsibility and ownership of risks. The Sarbanes-Oxley Act (2002) is similar to the Turnbull Report. The Act introduced highly significant legislative changes to financial practice and corporate governance regulation in the United States.

The Turnbull Report is not just about avoidance of risk. It is about effective risk management – determining the appropriate level of risk,

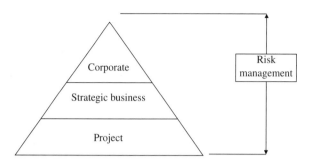

Figure 10.1 Levels within a typical corporate organisation.

being conscious of the risks taken and then deciding how to manage them. Risk is both positive and negative in nature. Effective risk management is as much about looking to make sure that you are not missing opportunities, as it is about ensuring you are not taking inappropriate risks. Some organisations will seek to be more risk averse than others. However, all should be seeking to achieve a balance between encouraging entrepreneurialism within their business and managing risks effectively.

The purpose of the Turnbull Report is to guide British business and help it to focus on risk management. Key elements of the report include the importance of internal control and risk management, maintenance of a sound system of internal control with the effectiveness being reviewed constantly, the board's view and statement on internal control, due diligence and the internal audit.

Figure 10.1 shows the levels of a typical organisation structure, which allows risk management to be focused at different levels. By classifying and categorising risk within these levels it is possible to drill-down or roll-up to any level of the organisational structure. This should establish which risks the project investment is most sensitive to, so that appropriate risk response strategies may be devised and implemented to benefit all stakeholders.

Risk management is seen to be inherent to each level although the flow of information from one level to another level is not necessarily on a top down or bottom-up basis (Merna 2003). The risks identified at each level are dependent on the information available at the time of the investment and each risk may be covered in more detail as more information becomes available.

10.2 Risk management

Risk management cannot simply be introduced to an organisation overnight. The Turnbull Report lists the following series of events that need to take place to embed risk management into the culture of an organisation:

- *Risk identification*. Identifying on a regular basis the risks that face an organisation. This may be done through workshops, interviews or questionnaires. How it is done is not important but actually carrying out this stage is critical.
- *Risk assessment/measurement*. Once risks have been identified it is important to gain an understanding of their size. This is often done on a semi-quantitative basis. Again, how is not important, but organisations should be measuring the likelihood of occurrence and impact both in terms of image and reputation as well as the financial impact.
- *Understanding how the risks are currently being managed*. It is important to profile how the risks are currently being managed and to determine whether or not this aligns with an organisation's risk management policy.
- *Reporting the risks*. Setting up reporting protocols and ensuring that people adhere to such protocols is critical in the process.
- *Monitor the risks*. Risks should be monitored to ensure that the critical ones are managed in the most effective way and the less critical ones do not become critical.
- *Maintain the risk profile*. The need to maintain an up-to-date profile in an organisation to ensure that decisions are made on the basis of complete information.

Often risk management forms part of an integrated management system along with quality management, planning, health and safety management and change management. In a competitive economy, profits are the result of successful risk taking. Against this background, the Turnbull Report, endorsed by the London stock exchange in the same year, strives not to be a burden to the corporate sector, but rather reflect good business practice. Organisations should implement any necessary changes in a way that reflects the needs of their business and takes account of their market. As and when companies make those changes, they should understand that they are improving their risk management and, consequently accrue an overall benefit that justifies any cost involved.

10.3 The risk management process

Figure 10.2 conceptualises the risk management process. Risk Management looks at risk and the management of risk from each organisational

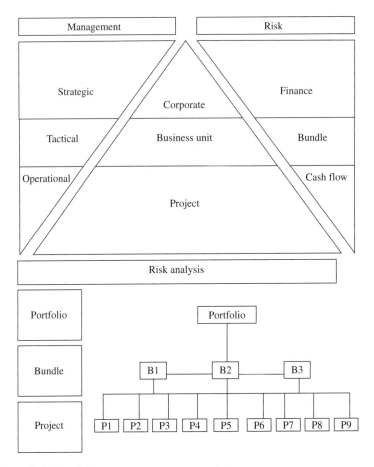

Figure 10.2 The risk management process and structure.

perspective, these being; strategic, tactical and operational. The level within an organisation responsible for each organisational perspective can perform the necessary analysis.

Organisations have different levels with different objectives. The risk management process separates the business processes into three levels, though many levels make up an organisation (these three levels being; corporate, business unit and project levels). Risks specific to each level are then identified using risk identification techniques and are logged on a risk register.

Each level within the organisation will then analyse the identified risks and responses and contingencies can be made.

The risks identified at each level are consolidated and controlled by a single department within the organisation. At this level within the risk management process, analysis can be made on either a stand-alone basis or for bundles of projects (portfolios).

10.4 Benefits of risk management

Risk management is not simply the identification of risks. It is about analysing the implications, response to minimise risk and the allocation of appropriate responses.

Risk management should be a continuous loop rather than a linear process, so that as investments or portfolios progress, a cycle of identification, analysis, control and reporting of risk is continually undertaken.

The benefits of risk management include:

- ❏ risks associated with the investments or portfolios are clearly defined well in advance of the venture;
- ❏ management decisions are supported through the analysis of data, allowing estimations to be made with greater confidence;
- ❏ improvement of investment or portfolio planning by asking What If questions with imaginative scenarios;
- ❏ the definition and structure of the investments are continually monitored;
- ❏ the provision of alternative plans, appropriate contingencies and concerning managers are part of a risk response;
- ❏ the generation of imaginative responses to risk;
- ❏ statistical profiles of historical risk being built up allowing improved modelling for future investments;
- ❏ investment issues are understood for each investment(s);
- ❏ decisions are supported by the analysis of data made available;
- ❏ the structure and definition of the investment or portfolio are continually and objectively monitored;
- ❏ contingency planning allows prompt, controlled and pre-evaluated response to risks which may materialise.

10.5 Recognising risks

For real-world organisations in viciously competitive environments, it is not good enough to simply protect the physical and financial assets of the corporation through a combination of good housekeeping and shrewd insurance and derivative buying. The pressure on margins is too intense and the vulnerability to volatility simply too great for it to be an adequate strategy for most organisations, even small ones. The focus therefore must shift to the far greater and far less tangible world of expectations and reputation, and thereby sustain investor value. Hence, the inexorable rise of risk management and its sudden popularity in the boardroom (Monbiot 2000).

Equity and credit analysts are increasingly focusing on risk and the quality of risk management within the organisations they analyse, which is further sharpening focus in the boardroom. Analysts want to be able to tell current and potential investors that the corporate management knows what it is doing and that it is using its capital in the most effective manner possible, and that it is in control of its strategic business units (SBUs) and consequently future profits.

Senior management are increasingly using company reports and press departments to boast about their latest risk management initiatives and policies, but learning the vocabulary associated with risk management and simply slipping the words into glossy brochures does not constitute risk management. Organisations that want to report the stable, secure, socially responsible and ever increasing earnings that investors and other stakeholders demand must take risk management seriously and put such words into practice (Merna 2003).

At corporate level, more enlightened senior management are hiring risk managers, more often than not originating from the insurance management and finance sectors. Typically these individuals are responsible for the identification, measurement and mitigation of risk, as well as arranging its funding when feasible and desirable. In many cases, these individuals have attempted to co-ordinate the risk management activities of other departments and to promote a risk management culture throughout the organisation. Lamb *et al.* (2000), however, noted that there were less than 50 designated risk officers worldwide, employed by internationally listed organisations.

A recent survey of CEOs and risk managers in the United Kingdom, Europe and the United States has shown that the main perceived risk issues today are – corporate governance, extortion, product tampering and terrorism, environmental liability, political risk, regulatory and legal risk, fraud and a whole host of risks ushered in by globalisation and modern technologies (Monbiot 2000).

The origin of these issues is many, varied and inextricably interrelated. But essentially, corporate and financial risk has grown in scale and complexity in tandem with the globalisation of the world economy. The globalisation of trade and investment and the removal of barriers at national and international levels have led to a massive process of consolidation in all sectors, as essentially uneconomic organisations, which previously relied on a combination of customer ignorance, lack of external competition and government assistance, have been forced to adapt or die.

In this global and increasingly service dominated economic environment, corporate success increasingly comes to rely on two key drivers – perception and knowledge. Risk management is an integral part of these and a thorough understanding of the concept will drive an organisation one step further to success. Organisations must have the ability to manage investments and ensure they are commercially viable. However, contingencies must be available, through the use of structured and upto date risk management systems.

One major risk to corporations is from hostile bids. Organisations, often increase their financial gearing to employ more debt than equity and thus make themselves less attractive to opportunistic takeovers. Shareholders, however, do not necessarily want too much debt as debt service is senior to dividend payment, which may result in poor or no dividends to shareholders.

10.6 Why risk management is used

Risk management can provide significant benefits in excess of the cost of performing it. Turner and Simister (2000) believe benefits gained from using risk management techniques serve not only the project or investment but also other parties such as the organisation as a whole and its customers. They suggest the main benefits of risk management are that:

❑ there is an increased understanding of the project, which in turn leads to the formulation of more realistic plans, in terms of cost estimates and timescales;
❑ it gives an increased understanding of the risks in a project and their possible impact, which can lead to the minimisation of risks for a party and/or the allocation of risks to the party best suited to handle them;
❑ there will be a better understanding of how risks in a project can lead to a more suitable type of contract;
❑ it will give an independent view of the project risks, which can help to justify decisions and enable more efficient and effective management of risks;
❑ it gives knowledge of the risks in projects that allow assessment of contingencies that actually reflect the risks and also tend to discourage the acceptance of financially unsound projects;
❑ it assists in the distinction between good luck and good management and bad luck and bad management.

According to Merna (2003) the beneficiaries from risk management include:

❏ corporate and senior management, for whom a knowledge of the risks attached to proposed strategic investment is important when considering the sanction of capital expenditure and capital budgets;
❏ strategic business managers, responsible for the risks attached to proposed tactical investment response;
❏ the project management team, who must meet project objectives such as cost, time and performance;
❏ the client, since this will reduce uncertainty in the overall investment outcome;
❏ stakeholders, uncertain in their particular involvement.

Risk management should be a continuous process over the whole life cycle of the investment.

Many project management procedures place considerable stress on the quantification of risk. However, at the strategic business and corporate level a significant proportions of the risks are not quantifiable and thus favour less formal risk management. The emphasis, placed on the quantification processes often leads to a failure at the corporate and strategic business level to prompt a manager to take account of other types of risk more difficult or impossible to quantify.

All stakeholder requirements must be acknowledged and aligned and a consensus must be found. This is often not easy, because stakeholders have conflicting interests. It is important that the positions of the stakeholders are continuously analysed and their expectations met as far as possible.

10.7 Model for risk management at corporate, strategic business and project levels

Within any organisation performing risk management, tools and techniques must be used at each level. The use of these tools and techniques allows identification and analysis of risks and forms the basis for investment appraisal. Stakeholders are also identified at each level, and are allowed to contribute to the risk management process. These stakeholders, must be identified and their requirements recorded as well as their relative significance. In order to assess the risks at each level, various tools and techniques may be applied. These techniques may generally be applied at each level in the process, but some will be more applicable to a particular level than others. Figure 10.3 illustrates the levels and required

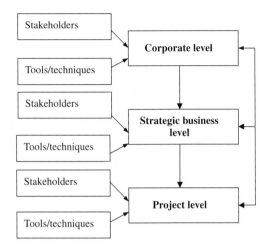

Figure 10.3 Risk management mechanism.

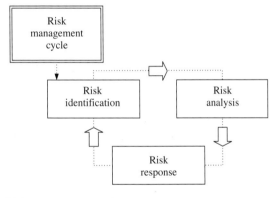

Figure 10.4 Risk management cycle.

input at each level in the risk management mechanism. The tools and techniques used at each level will be determined by the risk analyst and related to the type of assessment undertaken at those levels.

Figure 10.3 divides the organisation into corporate, strategic business and project level. At each level, risk management tools and techniques are used and stakeholder requirements are taken into consideration. This process forms a basis for the risk management mechanism.

Figure 10.4 illustrates the risk management cycle that includes the identification, analysis and control of risks to be applied at corporate, strategic business and project levels. The risk management cycle is dynamic and must be continuous over the project investment life cycle.

This risk management mechanism, proposed by the authors, illustrated in Figure 10.2 incorporates the risk management cycle shown in Figure 10.3 and is utilised at each organisational level.

At each level within the organisation, the authors propose a generic system, illustrated in Figure 10.5 with the purpose of identifying, analysing and responding to risks specific to each level within the organisation.

This process illustrated in Figure 10.4 should be a dynamic process carried out throughout the whole investment lifecycle in a continuous loop.

Figure 10.5 illustrates the processes, the authors suggest should be undertaken at each level of an organisation, the stakeholders and risk management tools and techniques being involved as and when appropriate.

The first step of risk management is investment appraisal at corporate level where the overall investment objectives are determined. It is imperative that the investment and derived objectives are identified and clearly understood at the strategic business level and by the project team. At this stage, each level of the organisation should define what the investment implications are at their level. For example, business or project requirements, client specification, work breakdown structure, cost estimates, project programme, cost and type of finance and project implementation plan. This is often performed through the use of historical data, organisational specific knowledge and from information specific to the project in hand and the organisation's overall goals.

The process of identifying risks is carried out through the use of a variety of techniques suited to the type of project and the resources available. The allocation of risk to owners is undertaken during this stage, which aims to allocate ownership of risk to the individual best placed to control and manage it. Identified risks and risk owners will be recorded on the risk register, which later will become a database at SBU level.

The information gathered at the identification stage is then analysed. Risk analysis tools and techniques, either qualitative or quantitative, are now employed to provide a thorough analysis of the risks specific to the project at each level within the organisation. Analysis may include defining the probabilities and impacts of risk and the sensitivity of the identified risks at each level.

After completion of the identification and analysis processes, response to these risks can be carried out. This part of the process, is exercised through the use of risk response methods and techniques. If the response decision is to mitigate the risks the costs of mitigation must be assessed and budgeted for accordingly. Retained risks at each level will be identified in the risk register and constantly reviewed.

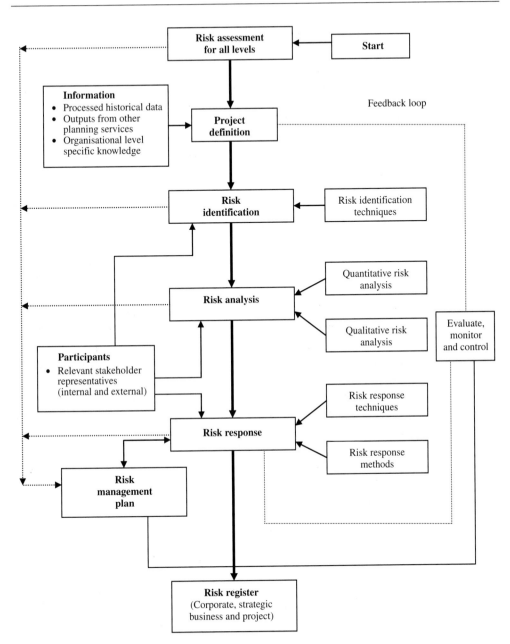

Figure 10.5 Risk assessment for all levels of an organisation.

Within this model, stakeholders are of particular importance. Stakeholders are involved at each level and will have an input at each stage in the risk assessment process (identification, analysis and response). The model allows information from each stage to flow backwards and forwards through the organisation, where it can then be continually monitored, evaluated and controlled.

Once all the information has been processed through the model, a risk management plan is constructed and implemented. The plan should form an integral part of project execution and should give consideration to resources, roles and responsibilities, tools and techniques, and deliverables. This plan will include a review of the risk register, monitoring of progress against risk, actions and reporting. The final output of the model being a corporate, strategic business and project level risk register.

Feedback is a key vehicle used in this proposed model for the organisation to learn from both its successes and mistakes, internally or externally. It provides continuous improvement at both SBU and project levels and risk management itself.

Feedback is a continous process gathering data from known and unforeseen events. Information is held at SBU level and disseminated throughout the organisation.

These corporate, strategic business and project level risk assessments and risk registers will be made available to each level within the organisation.

An overall risk register, incorporating risk registers developed at corporate, strategic business and project levels will be further developed at strategic business level and be continually updated as the project develops. It is important that the risk assessments carried out for the projects at SBU level are of the same format, thus providing a database for all projects. This will allow the database to be interrogated and inform future projects, strategic business and corporate decision making.

The authors suggest that risk assessment at corporate, strategic business and project levels should run concurrently. At any time during the assessments, risks can be flagged up from any level that may result in the project or investment not being sanctioned.

The proposed risk management assessment system will:

❏ identify and manage risks against defined objectives;
❏ support decision making under uncertainty;
❏ adjust strategy to respond to risk;
❏ maximise chances through a proactive approach;
❏ increase chances of project and business success;
❏ enhance communication and team spirit;
❏ focus management attention on the key drivers of change.

Figure 10.6 illustrates the risk management model and the interaction of each level within the organisation. Information regarding risk assessment and risk registers is passed freely through the organisation.

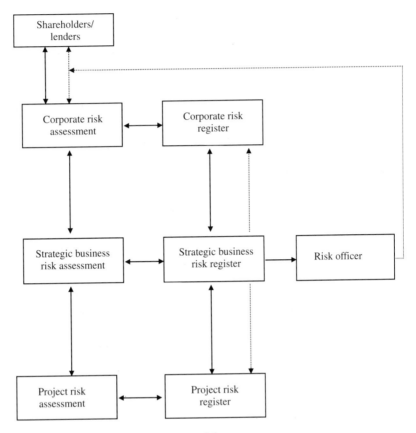

Figure 10.6 The risk management model.

Within this model, strategic business level will act as the conduit between corporate and project level. A risk officer, will be designated at the strategic business level with responsibilities for ensuring risks managed at corporate, strategic business and project levels are registered and that any further risks identified will be incorporated in the risk register held by the risk officer. All the information gathered from corporate, strategic business and project levels would be collated and passed on to the risk officer. The risk officer will be in direct contact with risk facilitators at both corporate and project levels. This model, will ensure that all levels of the organisation will have an input into the overall risk register.

Managers and owners of risks retained and mitigated will be in either the corporate, strategic business or project level within the organisation depending where the risk originates. For example, a risk originating at project level will be managed and owned by the project manager. The risk assessments and risk registers held by the project manager will be passed to the risk officer, at strategic business level. The risk officer will review the overall register and inform both corporate and strategic levels of any changes in risk assessment as the project proceeds.

The advantages of the strategic business level of an organisation holding a risk register as a conduit from both corporate and project level include:

- strategic business level is immediate to both corporate and project levels;
- one risk officer is responsible for the library of risk;
- if any information is required about risk specific to a project, both project and corporate levels will know where to access this information from;
- both project and corporate levels, will have access to all risk management systems and information;
- stakeholders, will have easy access to how risks are managed at all levels of an organisation;
- risk management throughout the organisation is co-ordinated and centralised.

However, in order for the model to work regular reviews and audits need to take place together with risk workshops at corporate, strategic business and project levels managed by the risk officer.

New risks, the cost of managing such risks and the status of all existing risks identified at each level will be addressed in the overall risk register held by the risk officer at strategic business level.

10.8 Summary

This chapter identifies the corporate, strategic business and project levels in a typical organisation. Each level, is responsible for managing the risks identified and ensuring that information on such risks is available to the other levels.

In most cases, risks are specific to each level. Corporate risks are typically difficult to quantify and manage. These risks include the political, legal, environmental and finance elements of an investment. Many of these risks can be assessed in greater detail at strategic business level as more information becomes available.

Project risk management, often entails risks being assessed in even greater detail as risks become more specific to the project rather than higher level risk discussed at strategic business and corporate level.

To ensure all risks at all levels are managed, it is paramount that an overall risk management system is implemented and risks identified at all levels are managed over the investment life cycle.

The risk register managed by the risk officer at strategic business level forms a database for all levels of the organisation. This risk register should be accessible to stakeholders, particularly shareholders involved in a project investment.

The continual cycle of risk management is fundamental to the risk management model illustrated in Figure 10.6.

References

Cadbury Report on Corporate Governance (1992) Prepared by Sir John Cadbury, Gee (a division of Professional Publishing Ltd.), London.

Lamb, C.W., Hair, J. and McDaniel, C. (2000) *Essentials of Marketing*, South-Western Publishing, Ohio.

Merna, T. (2003) *Appraisal, Risk and Uncertainty in Management and Corporate Risk*. Edited by N.J. Smith, Thomas, Telford, London.

Monbiot, G. (2000) *Captive State. The Corporate Takeover of Britain*, Pan Books, London.

Reichmann, P. (7 March 1999) Profile business, section 3, *Sunday Times*, p. 6.

Turnbull Report (September 1999) Internal control: guide for directors on the combined code, The Institute of Chartered Accountants, London.

Sarbanes-Oxley Act. (2002) Pub. L. No. 107–104, 116 stat. 745.

Turner, R. and Simister, J. (2000) *The Handbook of Project Management*, ISBN 0566081385, Gower, Aldershot.

Chapter 11
Case Studies

11.1 Introduction

The principal objective of this book is to provide practical advice and examples of project risk management applied to construction projects. Often books on this subject reiterate the theory without providing examples, which demonstrate how the techniques and tools can be applied in real situations to provide a basis for management decision making.

This chapter contains two case studies, both based on real projects but demonstrating different viewpoints and different techniques.

The first is related to a ship construction project, the second to a major transportation infrastructure project – the Channel Tunnel Rail Link (CTRL). The first is from the client's viewpoint of a major but conventional capital project, the second from the viewpoint of one of the consortia bidding for the 999-year concession the finance, design, construct and operate the high-speed trail link. The first demonstrates the use of influence diagramming methods, the second relies on more conventional planning and estimating methods.

Despite their differences, both case studies show how risks can be identified, assessed, modelled and analysed to provide information, which can be used by managers to assist in making decisions.

In both cases, changes were made to baseline programmes or estimates to improve confidence in the achievement of the projects objectives. In both cases, this required careful consideration of the risks in the programme – in all capital investment projects, time to reach the market and commence revenue-earning operation are critical.

In the first case, changes to the programme reduced the probability of overrun and increased the probability of earning revenue as early as possible. The second case was different in that the scale and complexity of the project coupled with the physical constraints, particularly relating to the tunnelling works, meant that room for manoeuvre was limited. This called for a more conventional approach, particularly for contractors, of assessing contingency levels to accommodate risks if they could not be avoided. In programme terms, avoidance of delay can equate to acceleration, so the assessment of possible delay and its consequential costs can be the equivalent of paying for accelerative measures to help safeguard opening dates.

The case studies, therefore demonstrate different approaches to dealing with similar problems but from different perspectives and using different methods. Both studies demonstrate that careful consideration of risk is a prerequisite of understanding fully the issues of uncertainty, which can affect almost all projects. Issues which, if they are not dealt with, will threaten the success of the project.

11.2 Case study – cruise ship design and fabrication programme risk assessment

Introduction

The case study describes parts of the risk management process applied on an investment project – the planning, fabrication and commissioning of a cruise vessel for American tourists in the Caribbean. The client was a large international cruise operator. The risk management included schedule, cost and profitability studies; however this example is limited to the schedule risk analysis.

The client was much focused on project risk, as a cruise vessel must be ready for operation at the start of the season. Any delays could cause dramatic effects for the reputation and, of course, the net present value of the project. Ticket sales would start 6 months ahead of the season, and all marketing material would be completed almost a year ahead. So, if the vessel arrived 4 months late, the first season would be spoiled, competitors could use the situation as an opportunity and the operator may lose customers to them.

Risk analysis was performed in the concept stage to assess the risk affecting project execution. The client wanted to verify the schedule to ensure that the probability of meeting the target date was high. They had selected a fabrication yard, and the project was about to enter the design phase. They knew that the most effective changes to the project could be done during the low cost design. Any changes made during the construction phase would be very expensive and could cause additional delays, the impact of which would be difficult to predict.

Main objective

The main objective of this study was to assess the probability of meeting the project completion milestone, to identify and quantify risks affecting the project schedule and to come up with changes to increase the probability if necessary. The study focused on:

❑ identifying the total risk exposure;
❑ addressing the main risk contributors;
❑ establishing a risk management plan; and
❑ initiating follow-up actions to reduce negative effects of risk and exploit opportunities.

Project key data

Project key data represent the information necessary to understand the project and to produce the model without going into too much detail.

Project duration

The fabrication of the vessel was estimated to take 16 months from its target start. This included startup activities, procurement, construction, installation testing and commissioning. The vessel had to be completed within 19 months, the time between project startup and start of the new season. It appeared therefore that the programme contained 3 months of float.

Limitations

This case study focuses on the main schedule for the fabrication of the vessel, and how strategic decisions based on results from the risk management process contributed to successful project completion. This example does not discuss in detail all information available, all the risk management processes or all details of the risk model.

11.3 Risk identification

In the risk identification phase, potential risks, their consequences and interrelationships were identified in creative workshops and from interviewing key personnel. During this process the following main risks affecting the schedule emerged:

❑ Innovative technical solutions could lead to uncertainty related to the duration of the design phase.

- Complex design and limited resources could lead to risk of a delay in the construction phase.
- The yard was already committed to another contract, which was due to start just after planned completion date of the cruise ship. This imposed a tight schedule on the cruise ship and could have lead to a potential conflict if the cruise ship was delayed.

The yard had been chosen for the new cruise ship because an option in a previous contract between the yard and the client offered a very good deal to the client. In other circumstances, the option of using another yard to avoid potential conflict if the project was delayed could have been explored.

Risk analysis

In the analysis phase the identified risks are quantified. A risk model was established using the DynRisk™ software (risk analysis tool based on the graphical influence diagram modelling technique and the Monte-Carlo simulation method).

Quantified risks are included in the risk model by estimating a pessimistic, a most likely and an optimistic value, together with probability distributions. The latter, can select from a wide choice of probability distributions, but it is recommended to keep it simple. All the information is very uncertain, and it is not recommended that a lot of time be spent on the probability distributions. It is much more important to try to open the minds of the project team during the identification process.

All results from simulations are available both in graphical and tabular format. The tool is very quick, which is an advantage when performing what ifs to assess the effect of introducing new variables or changing existing ones.

DynRisk™ enables the transfer of information gathered in the identification phase into a risk model reflecting the real risks affecting the schedule. This is very important. The software package used must not put limitations to the risk process. A graphical and flexible tool can also improve the modelling creativity and create a deeper insight into the relationships between the activities that make up the project and the risks that affect them.

The schedule

Based on information gathered in the identification phase, the following schedule model was established. This is a simplified programme using only

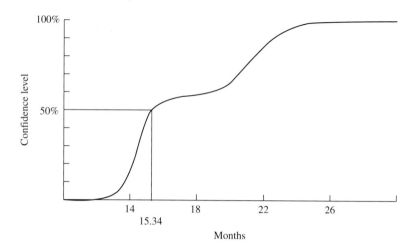

Figure 11.1 The initial schedule risk assessment.

the activities on the critical path, which shows only sufficient information to model the high-level risks.

The simplified programme could be modelled as an influence diagram. In this form, the model is easier to understand and communicate.

A Monte-Carlo simulation is performed. DynRisk™ generates different plots to visualise the results. The cumulative S-curve plot (Figure 11.1) reflects all possible outcomes of the schedule model, as follows. The ship had to be ready for operation within 19 months. The S-curve shows that there is a 50% chance of finishing the vessel within approximately 15 months. However, the step in the curve reflects a 40% chance of missing the first cruise season. Simulation of the overall model showed that the risks affecting the schedule also had a serious impact on the profitability of the project. If delays occurred, costs would increase, the season could be missed and the profitability targets would no longer be valid.

Further analyses were made to assess actions to change the project to increase the probability of meeting the completion date. The results showed that focus must be put on the design activity and the hull construction to reduce the probability of a delay.

Based on the information available and the results, an action plan was created. From the results of the analysis it is obvious that the client had to avoid a situation where no dock was available. Focus was put on trying to reduce the probability of a delay occurring. The following responses were feasible choices:

❑   *Re-schedule.* Design and construction of the hull proved to be the most critical activities as it was crucial to get the vessel into the dock

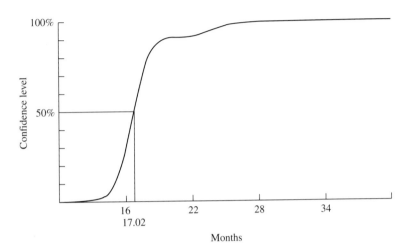

Figure 11.2 The adjusted schedule risk assessment.

on time. Hence these were the main risk contributors. A new production plan was established and more resources were dedicated to these activities. The design phase was modified to introduce a second phase in the hull design process.

❏ *Less innovative design.* The design phase was a main risk contributor, therefore compromises were made and more off-the-shelf solutions were chosen and some nice-to-have features were removed.

❏ *Buy time.* Negotiations led to an agreement between our client, the yard and the next client. The next client agreed to postpone their project for a considerable compensation if a delay occurred.

The responses were included in the new model (Figure 11.2). This adjusted model generated the following results:

The project now has a 50% chance of completion within 17 months, a two months increase from the previous scenario. The increase is due to less innovative design, that is, safe but time-consuming design and fabrication. The duration increases, while the risk is reduced.

The chance of delaying the project and failing to meet the completion target is now dramatically reduced. The client now has a 90% chance of meeting the target of completing the project within 19 months.

The client now has a much better basis for making decisions based on which actions to initiate. The client will probably not need to buy time to secure the dock availability because of the high probability of completing on time.

The probability of completing within 19 months is 90%. This may be too high. A more detailed analysis should be undertaken to refine this estimate. It can be very costly to reduce the risk too much because all risk reduction has a price. The client would probably look further into the design activities to further investigate the possibilities of a more novel design to try to save money, even if this reduced the confidence of completing within 19 months.

Conclusion

The client and the project team now feel much more confident that they will meet the target. The risk management process proved again that a schedule or a cost estimate with single number is unreliable. Most important, a risk management process is needed in every project phase to create a realistic basis for making decisions. In this example, the risk software tool, with focus on modelling flexibility and the graphical approach making it possible to communicate the model contents to all levels within the organisation, helped the risk management team. The effects of changes were tested on the model, and the results helped the team to come up with the best actions to secure the project against delays.

11.4 The Channel Tunnel Rail Link (CTRL)

This case study, outlines the risk assessments and analyses performed during the preparation of the proposal to design, construct and operate the CTRL by one of the bidding consortia (henceforth called the consortium). The consortium comprised contractors and consultants with financial and legal advisors. Banks were subsequently included to provide financial backing. The study focuses on the major risks identified during the preparation of the bid, how they were assessed and analysed and how a preliminary allocation of risk was made. Although this study demonstrates the application of the techniques of risk analysis and management to a mega-project, the approach is typical and can be applied to any size of project. Indeed the risk models and analyses are much simpler than may be adopted for smaller projects where the input data can be defined more precisely.

The objective of this exercise was to identify a realistic probable capital cost for the design, construction and commissioning of the infrastructure work including the stations, tunnels, track, signalling and control systems. This was necessary to provide information for the business case and to provide confidence to potential investors. It was important to give assurance that the capital cost estimate and programme were soundly based, realistic and achievable. This was particularly important insofar as the cost overrun and delays that affected the Channel Tunnel itself were still in the forefront of potential investors' minds.

Several of the bid team had been heavily involved in Eurotunnel's project management team. They were therefore acutely aware of the potential problems the rail link faced. One of the principal problems Eurotunnel had faced, was that of delay compounding cost increases. Cost and time are interdependent. Delays increase time-related costs of resources, magnify the effects of inflation and delay revenue earnings. For the rail link – and other design, build, finance and operate projects – this is particularly vital since the capital cost is only one factor in the business case, the other factor being the revenue streams that usually start only on completion of the capital investment phase. In this case, revenue generated by the Eurostar services existed from the outset and the extent to which this would pay for construction of the rail link heavily influenced the proposals. Nevertheless, as the Channel Tunnel demonstrated, the capital cost can be extremely important, particularly where it is sensitive to delay. The risk assessments therefore examined both the programme and capital cost; the former in greater detail.

11.5 Brief history of the CTRL

The desirability of a high-speed rail link between Britain and mainland Europe has long been recognised but its realisation depended upon the construction of both a cross-channel link and a high-speed rail line from a point within or close to London through Kent to the cross-channel link.

The cross-channel link has a long and turbulent history, culminating in the Channel Tunnel owned by Eurotunnel and including several previous unsuccessful schemes. The 1974 scheme was aborted after construction had started because the cost of constructing the rail link between Folkestone and London was deemed to be more than the government at that time could afford. The prime reason for this cost was increasing awareness that significant measures to protect the environment would be required.

Despite this, in the mid-1980s, potential promoters of the Channel Tunnel project were told that the high-speed rail link would be open at around the same time as the cross-channel link and Eurotunnel's business case made allowance for this. Again, however, the estimated cost of the rail link project, combined with the British Government's reluctance to subsidise it, meant that it was postponed, while the private sector put forward schemes. The intention was that the project was to be funded substantially by the private sector, although unlike the Channel Tunnel, it would receive a limited subsidy from the Government. The objective of the proposal that is the subject of this study was to produce the bid that required the lowest subsidy.

The Channel Tunnel was officially opened in May 1994, with revenue earning services building up slowly over the following year. In parallel with construction of the Channel Tunnel, the rail link underwent a prolonged gestation period. Several schemes were put forward by private sector consortia but these were not accepted. The final choice of route and the responsibility for preparing preliminary design and tender documents were given to a specially formed subsidiary of British Rail called Union Railways. Its design envisaged the high-speed link running from Folkestone to St Pancras via the newly constructed international station at Ashford, with a link to the international terminus at Waterloo station. The rationale for this scheme was that as demand for the Eurostar service increases, capacity of the terminus at Waterloo would be exceeded. Another London terminus will therefore be required. After consideration had been given to siting this at Stratford in east London, it was decided that a terminus closer to central London would be preferable and after King's Cross had been considered but rejected, St Pancras was chosen. This would ultimately allow construction of a major transport hub with links to the London underground system close to the main rail lines to the north of England, the Midlands and Scotland. The station at Stratford was to be retained to aid the economic regeneration of the East End of London. This option entailed tunnelling for approximately 17.3 km from the eastern edge of London's suburbs to St Pancras plus tunnelling under the River Thames. The completion date for the project was 31 December 2002.

Four private sector consortia were shortlisted to put forward detailed proposals. The consortia comprised contactors, consultants, design specialists, project managers, travel operators and bankers. Two of the consortia had never worked together before; the other two had been involved in earlier proposals. Several vanloads of tender documents were delivered to them in September 1994. The proposals were submitted in mid-March 1995. In this seven-month period the bidders had to examine Union Railways' outline design proposal, check its cost estimate, develop a programme to design, procure, construct and commission the link and propose a business case for the project that would require the lowest subsidy.

In the case of the bid that is the subject of this study, the team was composed of contractors, consultants and equipment suppliers supported by rail, finance and legal specialists. In order to complete the proposal in the seven-month period work proceeded on checking the estimate and development of the programme in parallel with the development of the business case and preparation of the financing and legal frameworks. The revenue side of the business case, was based around forecasting revenues from the Eurostar services. In addition, the bidders were free to propose commuter services from Kent to make use of the new rail lines. Any such services would necessitate the construction of additional stations. The bidders were required to submit a compliant bid, for the full scheme from St Pancras to Folkestone including the new international stations at Stratford and Ebbsfleet. The bidders also had to make allowance for providing enabling works for the Thameslink 2000 rail service by providing the structure for a new station below St Pancras with links to the existing London underground stations at St Pancras and King's Cross. The bidders' additional proposals had to be submitted as either alternatives or variants to the compliant bid.

Programme and constraints

Fundamental to the bid was understanding the programme outlined by Union Railways and in particular the constraints and milestones that are imposed. Around these, a detailed programme for design, procurement, construction and commissioning was developed. The programme constraints had a direct influence on the way in which the consortium planned its procurement strategy. Programme considerations also influenced considerations of contingency planning and allowances for delay and for acceleration costs.

The programme for the project was dictated to a large extent by the timetable for the parliamentary process. This involved drafting a hybrid bill, its passage through both Houses of Parliament and granting Royal Assent to it. The timetable for parliamentary process was as follows:

❏   Drafting and passage of the hybrid bill to be completed by late 1996;
❏   Royal Assent of the bill by early 1997.

The other significant date was the award of the concession following the bidding process. The award date was originally programmed for September 1995.

This timetable would enable the concessionaire to absorb Union Railways and inherit the existing European Passenger Services business consisting of the Eurostar high-speed trains, the existing international terminus at Waterloo and other infrastructure, including maintenance depots. The successful bidder would also inherit St Pancras railway station, which is a listed building and is to become the main terminal

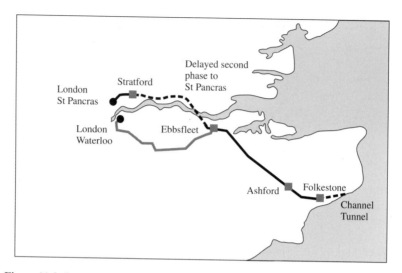

Figure 11.3 Proposed route of the high-speed rail link.

for the Eurostar service. The existing station was to be developed and extended without losing its Victorian façade or the view of the existing arched train-shed roof. Lands between the two existing mainline stations was also to be given to the successful bidder for development, though first the site of an old gas works had to be cleared and listed gas holders dismantled and removed.

Figure 11.3 shows the route of the line from St Pancras through tunnels under London and the new station at Stratford. The line was to emerge in east London, traverse Rainham Marshes through disused firing ranges, over the Dartford Tunnel, under the Queen Elizabeth II Bridge and under the Thames in tunnels. The track was to emerge near the new station at Ebbsfleet, then continue through a narrow corridor in urban areas across north Kent, across the River Medway, through the North Downs in another tunnel and across Ashford, eventually joining with the Continental Main Line at Eurotunnel's terminal near Folkestone.

Much of the route is through areas where Saxon and Roman remains can be found, through areas of special scientific interest, such as Boxley Wood or close to existing towns and villages. As a consequence of the latter, the route runs very close to the existing M20 through Kent. The main risks associated with these works are described briefly below.

11.6 The risk management process

The estimate of the capital cost of the CTRL and the associated implementation programme was the subject of risk analyses to provide estimates of the likely impact of risks not covered in the base estimates and pro-grammes. The purpose of the risk analyses was therefore clear, as were the team responsibilities. The data available were comprehensive and the procedure adopted followed three states:

❑ risk identification;
❑ risk analysis;
❑ response.

It is normal to break down the final stage into planning and managing. However, at tender stage the management actions are limited to quanti-fying the risks of the programme options and the procurement strategy for allocating risks.

The initial assessment concentrated on the proposed allocation of the risk between the parties involved in the project. This led to two forms of response:

❑ the first was to transfer risk to another party through a contractual and legal framework;
❑ the second was for the consortium to retain certain risks.

The risk assessment and analyses described in this case study were those required to quantify both the residuals of the transferred risks, such as the consequences of a contractor's bankruptcy and the risks, which were to be wholly retained by the consortium, such as delays to the overall implementation programme.

Risk identification

The risk management process commenced with a paper presentation to brief the bid team. Identification of risks commenced with the completion of pro forma sheets by specialist groups responsible for the construction proposals. These sheets identify the risk, the section of the project likely to be affected, the probability of the risk occurring and its potential impact on performance, cost and programme. A workshop was then held to review the major issues and risks.

Risk identification continued as work on the construction proposals, estimate, programme and business proposals proceeded. Ultimately all of the consortium's team contributed to the risk identification process and over 200 major risks were identified. Each risk was entered into a database that facilitated ranking and sorting by category and ownership. In parallel, the contractors' estimators undertook detailed assessments of the estimate prepared by Union Railways. Regular discussion and attendance at estimate review meetings assured that omissions and duplications were minimised. It was notable that the estimators tended not to identify some of the risks related to design development, planning consents, etc. because under a conventional contract the client or engineer would deal with these issues rather than the contractor.

The risks were allocated to seven categories:

(1)   project-wide uncertainties;
(2)   advanced and enabling works;
(3)   St Pancras terminus;
(4)   tunnels;
(5)   route sections (excluding tunnels);
(6)   intermediate stations;
(7)   system-wide mechanical and electrical equipment including track, power supply, signalling and data transmission systems.

The risks associated with each of these categories are described briefly below.

Project-wide risks

The proposed corridor for the track defined by Union Railways contained several sections, which did not appear to be viable in terms of the requirements for speed, locomotive power and track gradient.

Geotechnical information was incomplete. Additional surveys and boreholes were required, many of which would be needed before the Royal Assent of the enabling legislation, which entitles the concessionaire to have access to the sites. This meant that access for this work was dependent upon agreement of the existing landowners, many of whom were opposed to the project and likely to prevent the surveys taking place. This was of particular concern for the design of the London tunnels and for the design of the tunnel-boring machines (TBMs) because these have a long lead-time for design, manufacture and testing prior to delivery to site and erection. The criticality of these parts of the programme is discussed below.

Advanced and enabling works

These works needed to be carried out early in the project, some by Union Railways before the award of the concession, including major utilities and road diversions at St Pancras and along the routes. All the traffic flows had to be maintained during the works. The gasworks site at St Pancras had to be cleared and decontaminated. A major gas main had to be relocated, which British gas advised could be done only during the summer period. In all cases, planning consents were required. Any failure to progress this work would have significant impacts on the overall programme for the project and hence on its cost, by reducing the periods for design and construction of the main works.

St Pancras terminus

Seven significant risks associated with St Pancras were identified:

(1) resignalling and construction works whilst maintaining all existing rail services;
(2) relocation of services such as water, gas, electricity and telecommunications both within and adjacent to St Pancras;
(3) environmental measures including restrictions on working hours to reduce the nuisance to surrounding areas;
(4) heritage-related works including archaeological investigations and the recording and possible dismantling of listed buildings;
(5) construction of the belowground works and in particular, the Thameslink box around an existing underground line in a brick tunnel;
(6) complex phasing of road works and diversions and the programming of these works with the construction of the new station's east and west sides;
(7) design and construction of the roof of the new train-shed extension.

The main uncertainty common to all these risks is the role and influence of third parties during the planning and consulting stages of the design process and during construction. For example, the requirement to maintain the railway services including Thameslink (except for an extremely limited period) and the complex operational and constructional interfaces around the station that must be managed. These risks can be reduced by establishing early, close links between the interested parties, establishing good working relationships and by detailed planning, which incorporates the requirements of all interested parties. There are, however, significant and unresolved risks which, if they should occur, could jeopardise the programme and cost of these works. One particularly important consideration is that the design of the new train-shed extension required approval by the Royal Fine Arts Commission.

Tunnels

Six main risks affecting the tunnels were identified:

(1)   planning consents for the first site works that enable tunnel boring to commence at Stratford. This is extremely sensitive because of the need to commence very early with site surveys and enabling works;
(2)   tunnel design including diameter (being dependent on speeds, aerodynamics and space for mechanical and electrical equipment);
(3)   difficult ground conditions and the requirement to use sophisticated TBMs;
(4)   spoil removal from the London tunnels;
(5)   phased construction of a massive underground box for Stratford station;
(6)   construction of the eastern portal of the London tunnels at Barking.

The first risk is similar to the problems that will be encountered elsewhere on the project and especially at St Pancras, as noted previously. The others are related to the technical aspects of the works, including finalisation of the project's design and to the physical conditions that will be encountered during tunnelling. It was considered that these risks would be reduced by thorough design studies and through experience gained on other tunnelling projects that will be complete before work starts on this project. The removal, handling and transport of spoil required careful consideration to minimise the impact on the surrounding areas.

The risks related to Stratford station are those of constructing a major structure below ground in poor ground conditions. The critical impact is on the programme and the commencement of tunnelling. The ends of the box need to be constructed first, at the commencement of the construction programme, to enable the TBMs to be erected.

The final risk relates to the construction of the eastern portal of the London tunnels. Union Railways' design proposed that this should be located at Barking in east London, emerging in a suburb in close proximity to housing. Detailed analysis of the construction of the portal suggested that the constraints on working hours and the physical restraints due to the confined nature of the site would give rise to a significant risk that the portal would not be completed in time for the TBMs to emerge. It was therefore decided to allow for the cost of boring an additional 1 km of tunnels to emerge near Ripple Lane. At this location, construction was much simpler and quicker; the additional cost of the tunnels was more than offset by the easier portal construction and increased confidence in the programming of the works. After the proposals were submitted Union Railways changed the location of the portal to Rippleside.

Route sections (excluding tunnels)

Seven main risks were identified:

(1)   planning consents for works in Kent – again this was extremely sensitive;
(2)   extensive working alongside the existing railway and roads in a very narrow construction corridor;
(3)   wet weather delays to earthworks in chalk;
(4)   extensive requirements for aquifer protection;
(5)   a large number of bridges and other structures to be constructed simultaneously;
(6)   construction of the Medway crossing (an approximately £5 billion bridge);
(7)   construction of the North Downs tunnel.

The second risk presents a need to form working relationships with railtrack and with the highways authorities. Restricted working in a narrow corridor adjacent to the existing railway and roads requires close co-ordinator, detailed planning and co-operation at all levels of the organisations involved.

Major earthworks are required. These could be adversely affected by wet weather. The consortium made allowance for this and other risks related to the earthworks that were planned and estimated in great detail. The allowances for this were in the base estimate and programme. It was a requirement for the completed project that measures to protect aquifers along the route from contamination from the railway were put in place. As a result of the risk identification exercise, it was recognised that similar protection needed to be provided for construction haul roads and other temporary works. Allowances were included in the estimate for this.

Simultaneous with the earthworks, there are a large number of structures, including bridges and viaducts, to be built. This required careful planning both for co-ordination with railtrack, the statutory undertakers for electricity, gas, telecommunications, road authorities and also to ensure that specialist construction plant is available when required. Such is the extent of the work that the demand for cranage has the potential to exceed the numbers of cranes available in the east of England. The

Medway crossing and the construction of the North Downs tunnel need to be treated as major projects in their own right.

Intermediate stations

The intermediate stations did not have any detailed designs. It was impossible to say with precision what the risks would be. This in itself was a major risk. There are, however, risks associated with the proposed sites, which were identified. The major risks for Stratford are related to the construction of the box structure prior to the construction of the station.

The Ebbsfleet site presented several problems. These were:

❏ bad, marshy contaminated ground, with a high water table;
❏ potential archaeological remains;
❏ links to proposed new road.

Mechanical and electrical equipment, including signalling

The risks that could impact the design, procurement and installation are varied in nature and are depended on the equipment in question. In summary, the risks were:

❏ systems not fully defined, notably the signalling system, control centres and data transmission systems;
❏ installation of the equipment in the tunnels;
❏ track installation; and
❏ commissioning.

The lack of definition is of special importance for the signalling systems and control centres that are dependent upon decisions concerning the future operation of the railways. Development of bespoke software was required because proprietary systems were unsuitable. Lack of definition also affected the installation of equipment in the tunnels. Installation of track, including the availability of plant, logistics and planning, had to be carefully considered. It was recognised that there is not sufficient track-laying equipment in the United Kingdom to lay the entire track in the period available in the programme; hence allowance was made for importing this equipment. Finally, the risks to commissioning were to be minimised by considering testing, commissioning and operational requirements throughout the design, procurement and installation processes. Phased integration of operation's staff will be needed to ensure that there are no handover or start up problems caused by unfamiliarity with equipment or procedures.

11.7 Risk assessment, analysis and response

The completed risk register contained a diverse range of risks. These were assessed and analysed in three stages:

❑ First, risks were allocated on the basis of which party or parties should carry or share each risk. This led to proposals regarding the contractual arrangements for the project's management and execution that included provisions for the transfer of risk from the consortium or for sharing risks with other parties, including contractors and suppliers.

❑ Second, risks allocated to or shared by the consortium were reviewed and categorised according to whether or not they were included in the base estimates and programmes or in the project management and procurement proposals. That is to say that either allowances were included in the estimates and programmes or it was assumed that management action would be taken to avoid the risks. In the former case, the estimate and programme were risk adjusted and in the latter the analysis were based on risks being mitigated. It is essential for any quantification that a view is taken of the actions planned to reduce the risks and what effect these are likely to have on the impacts of the risks and hence on the values that are input to the quantitative analyses.

❑ Third, those risks retained by the consortium, either in whole or in part, were reviewed and categorised under four headings:

  (i)   The influence of market forces on the costs of resources and materials.

  (ii) The probability of cost increases due to increased scope, changes and variations, changing technology, etc. – covered by the term design growth.

  (iii) A specific uncertainty for the signalling, controls and communication systems, which are likely to be subject to the greatest technology changes and which were, at the time of preparing the submission, the least well-defined elements of the system.

  (iv) Programme delays resulting in cost increases either due to extensions of time, accelerated payments or a combination of the two.

These four categories were the subjects of risk modelling and analysis to determine their likely impact on the project.

The procurement strategy

The main mechanism for managing many of the risks was the procurement strategy. As noted earlier, the timing of the parliamentary process and the

Royal Assent of the enabling legislation governed the programme for the project. It was a requirement of the concession that the main financing package by banks, other institutions and individual shareholders must be in place within a limited period after the Royal Assent. To do this requires the preparation of a prospectus, which must contain first indications of cost that must be based on fixed-price tenders for clearly defined packages of work. However, the consortium considered that in the time available complete detailed design would be impossible. Its approach therefore was to prepare the designs in as much detail as possible then invite tenders on that basis. The tenderers would then price the completion of the detail design and the risk associated with that plus the construction, installation, testing and commissioning works. The outline designs would then be novated to the successful bidders.

There were two exceptions. First, the design of the signalling and associated systems was not sufficiently advanced for prices to be obtained. An estimate was to be put into the prospectus. Second, the TBMs for the London tunnels had to be ordered before the financing package was in place if the programme was to be met. That is to say, they were to be procured, at risk, by the consortium. A difficulty with this strategy was that the consortium contained several contractors, who expected to obtain significant parts of the work, which could have led to conflicts within the consortium and less than optimal prices.

Risk modelling

Two risk models were developed:

(1)   programme risk;
(2)   cost risk.

The cost risk analysis included the impact of the programme risk expressed in terms of the time-related costs, the calculation of which is described below. The development of the models and review of the results was an iterative process that fell into three phases:

(1)   Development of a cost model for a not to exceed estimate in December 1994. This was required so that preliminary discussions could be held with potential funding institutions.
(2)   Development of the programme risk models and review of the results.
(3)   Finalisation of the cost model based upon a much refined base estimate including a firmer view of the design growth risk and including the results of the programme risk analysis.

Programme risk models

The first assessment of potential delay to the project's implementation, and hence delay to the commencement of revenue earning operation, was made for incorporation in the not to exceed estimate. It was based on a review of the project programme's key stages and of the potential impact of the major risks to its achievement. A conservative view was taken to ensure that a worst case was used in the estimate, although it should be noted that certain risks were excluded. These are detailed below. The range was assessed as follows:

❑ *Minimum duration.* Completion, 3 months earlier than the contract completion date (i.e. September 2002).
❑ *Most likely duration.* Completion, 6 months late (i.e. June 2003).
❑ *Maximum duration.* Completion, 15 months delay to opening (i.e. March 2004).

The cost related to the delay was calculated by computing the time-related costs for each part of the project. For the construction activities, these were assumed to be the overhead/preliminaries costs. The quarterly time-related costs calculated for the not to exceed estimate was £30.79 million, resulting in a range of costs as follows:

❑ minimum: −£31 million;
❑ most likely: +£62 million;
❑ maximum: +£154 million.

It should be noted that this assumed the delay occurs at the peak of the construction effort and affects civil and building works, the mechanical and electrical (M&E) works and the project management. This was a reasonable assumption for a project of this nature where the civil and building works overlap with M&E design and procurement activities.

The cost model

Cost risk analyses were performed using a spreadsheet model and Monte-Carlo simulation. The final model is shown in Figure 11.4. It was a very simple model including the base estimates for the construction and project management plus allowances for the risk categories described above. The simple model has several advantages:

❑ It is easy to understand.
❑ It can be modified easily to test changes in the base estimate and assumptions about risk ranges.

Consortium cost risk analysis							Cost risk: March 1995				
Element description	Net cost	Prelims	Overheads	Gross cost	Duration (quarters)	Cost/ quarter	Distn type	Value			Risk mean
								Min.	Most likely	Max.	
St Pancras (inc. cont. £6.4 m)	364		115	479	19	6	S		479		479
Route (inc. cont. £25.7 m)	1694		519	2213	12	43	S		2213		2213
Electrification	117		52	169	8	7	S		169		169
Signalling/communications	143		41	184	8	5	S		184		184
Design				106			S		106		106
Sub-total	2318			3151					3151		
Project management, etc.	344			344	29	12	S		344		344
Sub-total				3495		73			3495		
Delay risk							T	0	145.6	218.4	121
Market risk (on gross cost inc. project management)							T	−174.75	0	349.5	58
Design growth risk (on £90 m allowance)							T	−4.5	0	9	1.5
EO for signalling, etc. (on £100 m)							T	0			25
Total				3495							3701.07

Triangular T
Single   S

Figure 11.4 Cost risk analysis.

❏ It avoids the need to correlate separate components of a complex model that do not behave independently. This is very important. Often models are too detailed and as a consequence produce misleading results.

The ranges of value applied to the market risk, design growth risk and the uncertainty related to the cost of signalling, control and communications systems were determined following review by the consortium's senior management and were as follows:

❏ *Market risk.* Minimum −5%, most likely 0%, maximum +10%. This range applied to the total base estimate including project management costs.
❏ *Design growth risk.* Minimum −5%, most likely 0%, maximum +10%. For the not to exceed estimate this range was applied to the total base estimate including project management. However, following the refinement of the base estimate the estimator's determined specific allowances for design growth on each section of the work. This allowance was £90 million contained in the base estimate. In the final cost risk model the range was applied only to this allowance rather than the total base capital cost estimate.
❏ *Signalling and control communication systems uncertainty.* It was assumed that a proportion of the estimated cost of these systems was at risk due to the lack of definition at this stage of the project, compounded by developing technology. The proportion at risk was assumed to be £100 million and the range applied was minimum 0%, most likely +25%, maximum +50%.

Finalisation of the cost estimate required a more objective assessment of the potential programme delay based on a more developed version of the consortium's own programme. An initial programme risk analysis was therefore undertaken based on a preliminary issue of this programme in bar chart format backed up by a network programme derived from the time/location analysis of the linear works along the route. The high-level summary programme in linked bar chart format was developed showing the main activities for the project's implementation, their interrelationships in terms of overlaps with lead and lag durations. This programme was reviewed by the specialist planning engineers responsible for each part of the project who revised the programme where necessary and determined the ranges to be applied in the risk analysis (Figure 11.5). Note that these are ranges of uncertainty. It is possible to separately model the key risks but as noted, it was decided to exclude these from this analysis.

ID	Name	Duration	Start	Finish	@RISK
1	CTRL and EPS Privatisation	261d	Jan 2 '95	Jan 1 '96	
5	Hybrid Bill Preparation and Support	460d	Jan 2 '95	Oct 4 '96	
3	Geotechnical and Site Investigations	130d	Jan 2 '95	Jun 30 '95	
70	Property Acquisitions	535d	Jan 2 '95	Jan 17 '97	Duration=RiskTRIANG (120, 130, 190)
71	Archaelogical Digs	369d	Oct 2 '95	Feb 27 '97	Duration=RiskTRIANG (525, 535, 601)
2	Engineering Design Development	455d	Jan 2 '95	Sep 27 '96	Duration=RiskTRIANG (358, 369, 457)
11	St Pancras - Adv. Works	1000d	Jan 2 '95	Oct 30 '98	
12	R1 St Pancras to Portals	954d	Aug 1 '95	Mar 28 '00	Duration=RiskTRIANG (756, 954, 1020)
13	R2 Barking Portal to Thames Tunnel	1000d	Jan 2 '95	Oct 30 '98	Duration=RiskTRIANG (978, 1000, 1110)
14	R3 Thames Tunnel to Medway	907d	Jan 2 '95	Jun 23 '98	Duration=RiskTRIANG (775, 907, 973)
15	R4 Medway to Bluebell Hill	628d	Jan 2 '95	May 28 '97	Duration=RiskTRIANG (584, 628, 760)
16	R5 Bluebell Hill to Charring Heath	632d	Feb 27 '96	Jul 29 '98	Duration=RiskTRIANG (566, 632, 674)
17	R6 Charring Heath to Mersham	840d	Jan 2 '95	Mar 20 '98	Duration=RiskTRIANG (774, 840, 906)
18	R7 Mersham to Dollands Moor	678d	Feb 27 '96	Oct 1 '98	Duration=RiskTRIANG (612, 678, 810)
19	Intermediate Stations	589d	Oct 2 '95	Jan 1 '98	Duration=RiskTRIANG (523, 589, 721)
21	R1 St Pancras to Portals	655d	Oct 2 '95	Apr 3 '98	Duration=RiskTRIANG (589, 655, 721)
22	R2 Modified Barking Port. to Th. Tunnel	783d	Oct 2 '95	Sep 30 '98	Duration=RiskTRIANG (717, 783, 849)
23	R3 Thames Tunnel to Medway	783d	Oct 2 '95	Sep 30 '98	Duration=RiskTRIANG (717, 783, 849)
24	R4 Medway to Bluebell Hill	717d	Oct 2 '95	Jun 30 '98	
25	R5 Bluebell Hill to Charring Heath	717d	Oct 2 '95	Jun 30 '98	
26	R6 Charring Heath to Mersham	935d	Oct 2 '95	Apr 30 '99	
27	R7 Mersham to Dollands Moor	849d	Oct 2 '95	Dec 31 '98	Duration=RiskTRIANG (783, 849, 915)
42	St.P. Roadworks Utilities	1085d	May 6 '96	Jun 30 '00	
43	St.P. Railway/LUL Works	1172d	Jun 5 '96	Nov 30 '00	
44	St.P. Train Shed Under Exist/New Deck	779d	Aug 30 '99	Aug 22 '02	
45	St.P. Train Shed	1090d	Mar 3 '97	May 4 '01	
46	St.P. Ancillary Building	1284d	Mar 31 '97	Feb 28 '02	
47	St.P. New East Side inc. Thames Link	264d	Jan 5 '98	Jan 7 '99	Duration=RiskTRIANG (242, 264, 308)
95	St.P. New West Side inc. Thames Link	407d	Jan 8 '99	Jul 31 '00	Duration=RiskTRIANG (385, 407, 451)
49	Thameslink Tunnels	631d	Sep 1 '97	Jan 31 '00	
93	Thames Tunnels (2 TBM)	0d	Aug 22 '02	Aug 22 '02	
51	Stratford Box (Short Box)	215d	Nov 4 '96	Aug 29 '97	Duration=RiskTRIANG (171, 215, 259)
52	Ebbsfleet Station	776d	May 9 '97	Apr 28 '00	Duration=RiskTRIANG (688, 776, 820)
39	TBM Specification and Procurement	544d	Oct 2 '95	Oct 30 '97	Duration=RiskTRIANG (500, 544, 610)
55	London Tunnels	610d	Sep 1 '97	Dec 31 '99	Duration=RiskTRIANG (488, 610, 758)
56	Thames Tunnels (2 TBM)	760d	Oct 7 '96	Sep 3 '99	Duration=RiskTRIANG (673, 760, 870)
57	North Down Tunnel	604d	Oct 7 '96	Jan 28 '99	Duration=RiskTRIANG (516, 604, 670)
58	R1 St Pancras to Portal	873d	Feb 28 '97	Jul 4 '00	Duration=RiskTRIANG (807, 873, 983)
59	R2 Modified Barking Port. to Th. Tunnel	808d	Aug 27 '97	Sep 29 '00	Duration=RiskTRIANG (742, 808, 874)
60	R3 Thames Tunnel to Medway	808d	Aug 27 '97	Sep 29 '00	Duration=RiskTRIANG (742, 808, 874)
61	R4 Medway to Bluebell Hill	743d	Aug 27 '97	Jun 30 '00	Duration=RiskTRIANG (677, 743, 809)
62	R5 Bluebell Hill to Charring Heath	743d	Aug 27 '97	Jun 30 '00	Duration=RiskTRIANG (677, 743, 809)
63	R6 Charring Heath to Mersham	849d	Jul 1 '97	Sep 29 '00	Duration=RiskTRIANG (783, 849, 981)
64	R7 Mersham to Dollands Moor	674d	Dec 2 '97	Jun 30 '00	Duration=RiskTRIANG (608, 674, 740)
92	Main Construction Complete	0d	Sep 29 '00	Sep 29 '00	
100	New Rolling Stock	534d	Dec 9 '96	Dec 24 '98	
28	Trackwork Design	320d	Sep 18 '95	Dec 6 '96	Duration=RiskTRIANG (319, 320, 452)
67	Trackwork Procurement	1049d	Dec 9 '96	Dec 14 '00	Duration=RiskTRIANG (1048, 1049, 1181)
97	Track Installation	390d	Nov 2 '99	Apr 30 '01	Duration=RiskTRIANG (389, 390, 456)

Figure 11.5 Sample of programme activity ranges.

The programme was developed on a simple proprietary planning package and analysed using a risk analysis software package.

11.8 Summary of the preliminary schedule risk analysis results

The results of the preliminary risk analysis of the base case project programme for all work post Royal Assent indicated that the project's completion date could be delayed to October 2003 compared to the desired opening date of 31 December 2002. On the basis of the assumptions made, this represents approximately 97% confidence based on the data used in the analysis. It should be noted, however, that a number of high-impact but low-probability risks such as finding significant archaeological remains were excluded (some were found near Folkestone at the end of 2000). This is an example of a risk that can be identified but for which no precise assessment of its impact or probability of occurrence can be made. Exclusion has the effect of reducing the potential delay and increasing confidence in the October date. The confidence level for October 2003 if these risks had been included would be reduced to, say, 80–85%. The precise duration and impact would depend on the time and location of the find.

Furthermore, it was assumed that Royal Assent would not be delayed and would take place in September 1996. It was also assumed that works such as geotechnical surveys, archaeological digs, property acquisitions and advanced works such as utilities diversions (which commence prior to Royal Assent) would not be delayed. As noted earlier, this was a sweeping assumption; it was likely that some of these works would be delayed, thereby reducing confidence in the October 2003 date still further.

A second analysis was performed that confirmed that a delay in the geotechnical surveys would extend the maximum completion date because the critical path ran through the tunnels construction activities preceded by design and procurement of the TBMs and the surveys. The critical path also runs through the installation of M&E equipment and through the testing and commissioning works. A Monte-Carlo simulation was used to combine the data and derive the output dates and a range for the project completion date. Subsequent development of the consortium's programme allowed risk analysis of a more sophisticated high-level network programme for the whole project. It concentrated on the activities that network analysis had indicated were critical, plus other activities that have the potential to cause significant delay.

The programme included dates for the completion of both the overall project and the early opening of the route to Waterloo. This was because the risk assessment suggested that the tunnels under the Thames and London, plus the route sections north of the Thames and work at St Pancras were the activities that were most at risk.

Two analyses were performed, therefore, to determine the risk of delay to both of these dates. The early opening to Waterloo could be achieved a year earlier than the total project but increased the criticality of the linear route works south of the junction to Waterloo and of the design, construction and commissioning of the control centre at Swanscombe. Since the signalling systems are one of the highest risk parts of the project, it was doubtful that the early opening of Waterloo would be viable.

The network programme was subjected to Monte-Carlo analysis. The results confirm that the lateness calculated at 97.5% confidence is approximately 9 months. The most likely completion corresponding to 50% confidence is approximately 6 months late. Again several low-probability, high-impact risks were excluded from these analyses, implying that real confidence was lower than computed.

Finally, the results of both the initial programme risk analysis and of the high-level network give comparable results. In addition, both of the analyses give similar results to the analysis published by Union Railways that indicated an opening date for the overall project of October 2003, calculated at 97.5% confidence (excluding catastrophic risks).

The closeness of these results indicated that based on the information available and making reasonable assumptions about the potential risks to the programme, the most likely (50%) opening date appeared to be between 4 and 6 months late and the 97.5% opening date between 9 and 10 months late. Again it must be emphasised that this excludes catastrophic high-impact but low-probability risks. There was a very low probability of achieving the target opening date for the total project.

The final cost model

The results of the final cost risk analysis were presented in three formats:

(1)   tabular summary report;
(2)   cumulative (S) curve (Figure 11.6); and
(3)   histogram of probability density.

The analysis indicated that approximately £573 million should be added to the base estimate, giving a total estimated capital cost of around £4 billion, corresponding to approximately 95% confidence (Figure 11.6). However, this was potentially optimistic since the potential for delay was

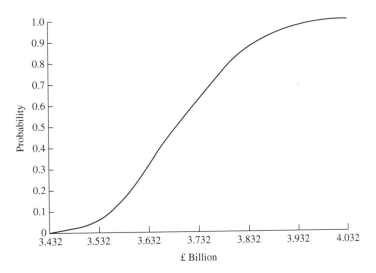

Figure 11.6 Simulation results.

greater than the analyses indicated because the high-impact risks had been excluded and potentially optimistic assumptions had been made.

Update

Following the submission of proposals the concession was awarded to London and Continental Railways (LCR) at the end of February 1996, several months later than planned. The main stakeholders in LCR were consultants, project management specialists, transport company operators, a power supplier and a bank. The group did not contain any contractors. This was a deliberate decision based upon Eurotunnel's experience where contractors had formed the original consortium and no strong independent owner and project manager had existed. LCR's approach emphasised the importance of developing the Eurostar business separate from designing and managing the project. The result was that LCR stated that it required a subsidy of £1.4 billion, reported to be £0.5 billion lower than the next lowest proposal. It was also to be paid much later during the construction phase of the project than requested by the other bidders. LCR has not published its capital cost estimate but figures close to Union Railways' £3 billion estimate were reported. Subsequently one of its senior figures suggested that funding of approximately £5 billion would be required.

In January 1998 LCR announced that it could not secure financing for the project, primarily because the revenue from the Eurostar services was significantly lower than projected. LCR requested a subsidy of a further £1.2 billion but the Government rejected its request. LCR were then allowed several months to submit a revised proposal.

On 4 July 1998, it was announced that agreement had been reached. The Government would not increase its subsidy but would underwrite the project's funding and Eurostar's losses in return for a share of Eurostar's future cashflows after 2020, assuming it becomes profitable, and a reduction in the concession period from 99 years to 90 years. The project will be split into two phases. The first phase will be from Folkestone to Ebbsfleet, including the international station there, and will be built first at an estimated cost of between £1.5 and 1.6 billion and was completed on time and budget in September 2003. It was intended that railtrack would purchase it for the cost of construction but following the Hatfield rail crash and the massive disruption to rail services as broken rails were identified and replaced, railtrack was forced into receivership. This was a risk that cold not have been foreseen – or if foreseen would have been assessed as very low probability.

The second stage from Ebbsfleet to St Pancras, including the station work there and at Stratford, commended in 2001 for completion in 2007 at an estimate cost of between £1.8 and 2.7 billion, giving a combined cost of between £4.2 and 4.9 billion at 1997 values.

Construction is proceeding largely on programme despite the collapse of one of the London tunnels under Stratford early in 2003 – luckily without loss of life or serious damage to property. The collapse was attributed to sunk holes caused by unidentified disused wells.

Chapter 12

Guidance in Practical Risk Management

This chapter integrates much of the theory, technique and practice outlined in the preceding chapters and offers guidance on how the process should be undertaken and executed in order to provide the project manager or client with meaningful, repeatable, supportable and useful information on which future decisions can be based. This guidance does not cover all viable approaches nor is it fully exclusive or exhaustive in the methods which are outlined; nevertheless, it will provide a useful support framework, particularly for the person undertaking this work for the first time. It will demonstrate the confidence which can be placed in the outputs and will reinforce the understanding of the nature of the project which can be obtained through the process of risk management.

12.1 Decision making

Initiating and managing construction projects requires decisions to be made. Risk management is a fundamental technique used to facilitate the decision-making process. Often decisions have to be made without a complete understanding of the precise basis upon which the decision is to be made, the reasons why it is being made or the consequences which flow from it being made. It may not always be possible to have complete information; however, it is important to recognise that the knowledge is incomplete and to try to assess the extent of the gaps and the potential consequences of taking action.

It is necessary to distinguish between the extent of knowledge and the impact of a decision. Two common problems can arise. First a tendency to delay making a decision, and second a tendency to seek more knowledge on the assumption that this will facilitate decision making. Seeking more information usually delays the decision-making process further.

Experience shows that introducing delay into a construction project is often, though not always, detrimental. It is therefore helpful to distinguish

Table 12.1   Decision classification.

Category	Extent of knowledge	Impact
Self-evident decision	Much knowledge	High impact
Simple decision	Much knowledge	Low impact
Arbitrary decision	Little knowledge	Low impact
Risk decision	Little knowledge	High impact

between types of decisions. One example of a decision classification is as shown in Table 12.1.

Obtaining more knowledge may serve to move some decisions into the self-evident category. However, it is the purpose of risk management to try to clarify potential risk sources and impacts even if there is never likely to be sufficient knowledge to make a decision self-evident. This will permit the risks associated with a particular project option or course of action to be identified and assessed in advance of decisions being made. This does not guarantee that the decisions would be better than a decision made under conditions of complete information but it would ensure that major risks are not overlooked, even if the decision may still be to proceed with the project.

Obviously, it is the last category, the risk decision that is most important and to which the majority of this book is dedicated. This decision is the most difficult and yet also the most significant. Improvements in project management depend upon improvements in the understanding, appreciation and execution of this decision. This book is aimed at assisting readers in this process.

12.2   Preparation for risk management

The obvious first question has got to be 'for which projects do we need to perform a risk analysis?' Unfortunately, there is no simple answer to this question. However, there are a number of project characteristics which, if present, influence the need for risk management procedures. For some organisations, given the combination of horror stories – for example, long-established companies going into liquidation due to the occurrence of unforeseen risks on a single project – and the improved access to risk management techniques, the question is turned around to read 'for which projects do we not have a need to perform a risk analysis?'

This question is easier to address and is also compatible with the concept of a hierarchical approach outlined in the earlier chapters. The starting point would be that all projects should be considered when this question is raised. It may be that there is one or a small number of simple repetitive, straightforward, fully controllable projects, undertaken by trained workers, with appropriate equipment, in a safe environment with guaranteed supply of raw material and guaranteed off-take or utilisation of product and no onerous time, cost or quality criteria to meet. Should any of these projects exist, then it may not be cost-effective to do anything further; however, for all other projects the first stage of a risk management process would begin. Broadly, risk management consists of potential risk source identification, risk impact assessment and analysis, and a managerial response to the risk in the context of the project. There are a large number of variations on this general theme but the one thing they all have in common is that risk must be managed in a systematic way via a number of stages, although the process should be regarded as iterative and the phases are not always sequential.

The scope of the project and the plan will be modified and changed as the risk management process progresses and it may also vary due to other external factors which in turn may require changes in the identification or assessment phases. Usually a top-down approach is adopted and the project objectives are clearly defined, sometimes with the aid of the early stages of the risk management process itself.

Once the objectives are known, there are a few simple questions which can be asked, regardless of the size, location, novelty or complexity of the project; these will assist in identifying the riskier projects. These questions might include the following:

- Is the client's business or economy sensitive to the outcome of the project in terms of the performance and quality of its product, capital cost and timely completion?
- Does the project require new technology or the development of existing technology?
- Does the project require novel methods?
- Is the project large and/or extremely complex?
- Is there an extreme time constraint?
- Are the parties involved sufficiently experienced?
- Is the project sensitive to regulatory changes?
- Is the project in a developing country?

Together these questions help to identify any projects which should definitely not be undertaken by the parties and those which, although risky, should be examined further by a rigorous identification of potential risk sources.

12.3 Risk identification

As stated earlier in the book, the identification process is concerned with risk sources and not with risk effects. Broadly, three differing methodologies were suggested: brainstorming sessions along lines similar to value management workshops, analysis of historical data for similar projects, and use of industrial checklists.

It is not possible to identify all possible risks, except in such a general manner as to be of little use. Nor is it possible to know whether all risks have been identified; but that is not the purpose of risk source identification. Again it should be stressed that perfect predictions of the future is not the goal of risk identification, rather it is the recognition of potential sources of risk for our particular project which are likely to have a high impact on the project and a high probability of occurrence. These are filtered out of a longer list of risk sources derived from the available data sources, people in workshops, historical data and advisory checklists.

So far, the most preferable method of identifying risk is the use of brainstorming, or similar techniques, which focuses each project team member on the risks specific to the project. The process must be carefully managed to remove individual and group biases as described in Chapter 3.

There is also the danger that the group does not have sufficient collective experience to identify all the key risks, even in a general form. This is why it is common practice to use external consultants or facilitators to prompt and guide sessions to produce a better balance assessment of project risk sources.

These potential sources of risk will form the framework against which the relative riskiness of various project options can be assessed. To do this, some form of quantitative analysis is usually undertaken.

12.4 Risk analysis

There are many methods of analysis, which require different levels of project knowledge and different data. These can range from the ranking of risks, which gives their relative importance but no quantifiable value, through to pseudo-quantitative techniques which introduce time or other parameters, to full simulation methods which provide ranges of programme durations, costs and rates of return. Not surprisingly, different methods will give answers in different formats but the inherent level of actual risk associated with a real project is the same whichever method is used.

This causes problems for some analysts and managers who are accustomed to receiving a single correct answer, irrespective of the method used. The key principle is that all methods of analysis give answers which reflect the inherent riskiness of the project in relative terms. Hence, if different methods of analysis are used, answers which appear to be different should be expected. It is important to note that the choice of method, or methods, to be used should be governed by the appropriateness to the project and the circumstances at the time of undertaking the risk analysis.

This book is not based upon one particular method; indeed it is not a question of deciding which method must be used and following this blindly. First, the hierarchical structure should be considered. Simple and rapid methods of risk analysis should be undertaken as a first step, only progressing to more complex, time-consuming and expensive methods as necessary.

However, if major risks are present in the project then it is likely that a full computer-based probabilistic analysis should be undertaken, if the impacts of the risks can be quantified. There are a number of methodologies for this but in this book the network-based, or influence diagram based, Monte-Carlo simulation has been recommended as the preferred method. However, it should be remembered that depending on the particular project, type of analysis most appropriate should be chosen.

After running the software analysis package, some analysts and text books seem to regard the process as complete; however, as has been discussed earlier, now this is not a widely held view. The process of translating computer software output into viable project decisions is a significant step in the risk management of projects and is too often neglected by practitioners. The following section describes how the outputs are used to provide information for the decision-making process.

12.5 Risk outputs

This section of the book examines the types of output which are produced by computer-based risk analysis packages and describes how to apply them to, explaining the key features and report options, communicating these findings in an appropriate form and considering their use in decision making.

Computer packages using a Monte-Carlo simulation will produce results in tabular and graphical format; usually the latter is preferable. Typically, three graphs are of interest: risk exposure, downside risks and risk contributors within each of the main project areas.

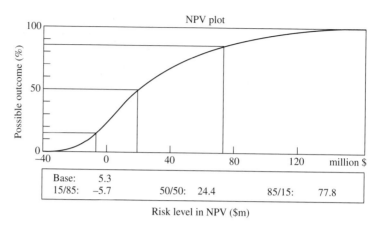

Figure 12.1 Risk exposure diagram.

Risk exposure diagrams

The project's risk exposure is the most important indicator of the project's riskiness. Risk exposure is usually illustrated by an S-curve, showing possible outcomes from 0% to 100% (sometimes shown as 0–1) along the *y* axis and risk level as measured by a project variable (in the case of Figure 12.1, net present value) along the *x* axis.

Before examining this type of diagram it is essential to understand that these diagrams will not always show the project's total risk exposure. There are three reasons for this:

❑   some risks cannot be modelled using risk analysis software;
❑   some risks may have been omitted from the model;
❑   some risks which are of low probability and do not influence the output greatly might have very serious consequences.

It is useful to consider Figure 12.1 carefully to understand what is being shown. Many analysts like to make a quick check on the 50% outturn, also known as the 50/50 estimate, which in Figure 12.1 is $24.4 million. By finding the zero point on the *x* axis the probability of a negative NPV can be found, in this case almost 20%.

Most useful is the range of likely outcomes that can be obtained from the figure. These values are not deterministic predictions of the likely performance of the project. The range, which is a function of the gradient of the S-curve, is a direct measure of the inherent riskiness of the project modelled and can be used to compare with other project options. The range taken for measurement is also the basis of discussion. Some analysts take the range from 15% to 85%. In this case that would equate to a pessimistic

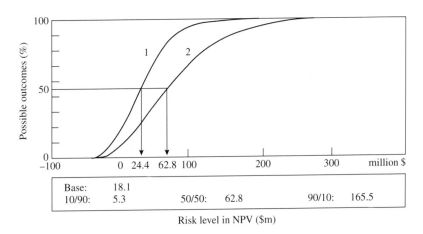

Figure 12.4 Revised risk exposure diagram.

context than is the norm. Examples of case studies, both hypothetical and real are used to demonstrate important principles and to generate risk outputs for discussion and review. The use of risk management undoubtedly brings many benefits to the construction project manager; however, unless conducted rigorously it can become stale and ineffective and in the worst cases reactive rather than proactive.

The purpose of the risk management process is to make effective project management decisions about what happens on the project tomorrow. It has to focus on the future, because future is the only dimension in which we can make effective change; yesterday has already happened and today things are in progress – so we must concentrate on actions and decisions which affect things from now onwards until the termination of the project.

The book is aimed both at undergraduate and postgraduate students and at the increasing numbers of engineers, surveyors and other professionals who are being required to study risk analysis during university courses and to develop this further through their professional practice. The needs at the practical level are significantly different from the needs at the theoretical level, and by isolating itself from detailed mathematical procedures the book concentrates on the provision of assistance with the execution of a practical risk analysis.

This book is a companion volume to the earlier Blackwell Science publication *Engineering Project Management* and the processes of risk management outlined here are fully compatible with the recommended project management philosophy and procedures.

Index

Page numbers *in italics* refer to illustrations